4

ESSAYS FOR THE THIRD CENTURY
America and a Changing World

Arms Transfers under Nixon

A POLICY ANALYSIS

Lewis Sorley

THE UNIVERSITY PRESS OF KENTUCKY

Scholarly publisher for the Commonwealth,
serving Berea College, Centre College of Kentucky,
Eastern Kentucky University, The Filson Club,
Georgetown College, Kentucky Historical Society,
Kentucky State University, Morehead State University,
Murray State University, Northern Kentucky University,
Transylvania University, University of Kentucky,
University of Louisville, and Western Kentucky University.

Editorial and Sales Offices: Lexington, Kentucky 40506–0024

Library of Congress Cataloging in Publication Data
Sorley, Lewis, 1934–
    Arms transfers under Nixon.

    Bibliography: p.
    1. Military assistance, American—History—20th
century.   2. United States—Foreign relations, 1969–
1974.   3. Nixon, Richard M. (Richard Milhous),
1913–   I. Title.
UA12.S67   1983   355'.032'0973   82–15970
ISBN 0–8131–0404–1

To the memory of my father

Contents

Foreword

The grant and sale of conventional arms have been, and will remain, among the most important instruments of policy in the politics of nations since World War II. Yet, compared to military strategy or economic aid, arms transfers have received very little systematic analysis, as opposed to polemical attention. Notwithstanding the general concern—especially in the United States—to curb conventional arms transfers (increasingly, in the form of sales), they have continued to serve the interests of the United States and other major (and some lesser) industrial-military states. Suppliers show little interest in collective restraints; recipients show no interest in self-restraint. But a growing number of states have shown a mounting interest in purchasing arms.

The reasons for the ubiquitous phenomenon of arms transfers lie not in the commercial interests of arms manufacturers or the balance-of-trade interests of governments—although, of course, both exist—but rather in the simple, yet complex, fact that arms transfers serve important political interests of both suppliers and recipients. They do so in a world in which other instruments of security, strength, and influence have relatively declined in utility. Arms transfers have not always served these interests well, but the same can be said of any instrument of policy.

It is important, therefore, for analysts of this phenomenon to understand the political functions of arms transfers and the conditions that underlie these functions. This applies equally to those interested in restraining arms transfers and to those interested in using them most effectively. Indeed, there is no necessary contradiction between the two objectives.

With this broad approach Lewis Sorley examines the role that conventional arms transfers played in the foreign policies of Nixon and Kissinger. The breadth and sophistication of his analysis make this examination more than a study of one important instrument of policy. Not only does it illuminate a particularly important period of American postwar diplomacy. It also tells us much about the dynamics of postwar international politics.

ROBERT E. OSGOOD

Preface

In the kind of retrospective policy assessment that here concerns us, the principle task must be not only to explain the structure of the problem, but also to describe in pertinent detail the context within which the policy was formulated and implemented. Context is particularly important in determining the utility, feasibility, and applicability of arms transfer policy during the Nixon era, for arms transfers represented nothing less than an attempt to salvage some influence, flexibility, and freedom of action for the president and his administration during an interlude in which the dominant forces were committed to constraining, diminishing, even undermining their ability to act.

Concern with arms transfers did not begin with the Nixon administration, of course, and in fact the latter years of the Johnson era were marked by active inquiry into and involvement in arms transfer decisions and policy on the part of the Congress. Beginning during that period and continuing through the Nixon years there were also major changes in the international traffic in conventional armaments.[1] These included a sharp increase in the worldwide trade as broadly defined, more suppliers of armaments, more recipients, a dramatic shift from grants to sales, greater sophistication of the arms involved in trade, diminished feasibility for all but the most industrially advanced states of maintaining a completely independent arms manufacturing capability, and continued reliance upon arms transfers as quid pro quo for various objectives of supplier states.

In their totality these changes resulted in a markedly different international environment. It became possible to a substantial degree for nations to purchase, rather than having to produce or be given, the military wherewithal for a modern armed force. The phenomenon carried with it important implications for dependency and obligation. It also raised many questions having to do with the strategic, economic, cultural, and ethical impact and utility of arms transfers. These questions were necessarily central to the foreign and defense policies of the Nixon administration.

Our analysis of the resulting arms transfer policy will begin with a description of the inherited situation with regard to arms transfers that confronted the Nixon administration when it came into office. Next we shall outline the larger context within which arms transfer policy was formulated and had to be implemented. Then we shall document the essentials of the stated policy and examine the most significant instances of implementation of the Nixon administration's arms transfer policy and the outcomes, in particular the case of the Middle East. This will enable us to formulate some general conclusions as to the adequacy, utility, and effects of that policy.

The research upon which those conclusions are based has included examination of the public record with regard to stated policies, analysis of detailed documentation with respect to policy implementation, survey of the periodical press for contemporary reaction to the evolving policy, a range of interviews with persons involved in formulating and executing policy at several stages of the administration's tenure, and consideration of the scholarly press and of industry journals for various perspectives on the material involved. A selected bibliography is provided which suggests the range of the resources consulted.

Although the focus of the analysis is on the policies devised by President Nixon and implemented during his presidency, in many respects the entire period from his inauguration through the truncated second term and the subsequent service of President Ford has been treated as an entity. So far as arms transfers are concerned, this approach is based on recognition that under President Ford there was a high degree of policy continuity, with essentially the same people continuing to make and carry out arms transfer decisions as under President Nixon. The approach and objectives continued to be consistent with those established early in his first term. Thus we have continued to speak of the Nixon policy even when referring to events that evolved or continued through the end of Mr. Ford's service.

The arms transferred to South Vietnam and other states in Southeast Asia during the course of the war there have been excluded from this analysis. They represented, in my view, a phenomenon wholly different from arms provided elsewhere in the world. Assistance to a wartime ally is not entirely a matter of volition, for upon joint success depends the outcome of the enterprise. To the extent an ill-supplied or ill-equipped ally hampers progress in winning the war, it is an integral part of one's own strategy and military effort to provide the necessary wherewithal. Even after United States involvement in Vietnam had ended, the provision of military assistance retained this character of aid within the context of a wartime collaboration and is thus not considered in this assessment.

At the Appendix may be found a set of figures and tables illustrating some of the key transfers of weapons systems and dollar value analyses of the flow and direction of arms transfers during the period under consideration.

As a civil servant, the author provides the usual assurances that the views presented are his own and that they do not necessarily represent those of any government department or agency.

Acknowledgments

The research upon which this work is based was undertaken under the auspices of the School of Advanced International Studies of the Johns Hopkins University, which provided an ideal atmosphere in which to pursue my academic and professional interests. I would like to thank especially Dr. Robert E. Osgood, then dean of the school, who was a friend and source of encouragement as well as a stimulating and perceptive critic.

Useful assistance was rendered by the staffs of the Army War College and Central Intelligence Agency libraries, and particularly the invaluable press archives of the Department of Defense, where Brant Keller and Joe Martinez were especially helpful. Unique research materials were provided by the Defense Security Assistance Agency, where the chief of the data management division, Stan Stack, and Fran Galliher were unfailingly helpful and cooperative. The editors and technical staff of the University Press of Kentucky have been a delight to work with.

Special thanks are due the members of my family. Kathy, Doug, Tim and Susan have been the most supportive of offspring, all while pursuing studies of their own. And my wife Ginny, a talented reference librarian, knows that in this as in all our shared endeavors I am forever grateful for her warm, cheerful, and sustaining spirit.

1. The Inherited Situation

Economic and military assistance as an instrument of policy has long been an important and, given the success of the Marshall Plan and the Truman Doctrine in the years following World War II, honorable adjunct of American involvement in world affairs. The idea of helping friendly nations to help themselves, especially where in so doing they were enabled to contribute to blocking the spread of hostile influence, appealed to both the prudence and the generosity in the national character.

While military sales were slender in the early postwar days of impoverished allies trying to rebuild war-damaged industrial bases and reestablish disrupted patterns of international trade, huge amounts of economic and military assistance were readily provided. Even where sales were concerned, however, it is instructive to note that, while foreign military sales agreements totaling some $11.3 billion were concluded worldwide by the United States during TY1950–FY1967, only about $5.6 billion of these, or slightly under half that total, were ever actually delivered.[1] That may be useful to recall in the present instance, where a great deal of attention has been focused on multi-billion-dollar agreements providing for deliveries stretching years into the future, only a small part of which has thus far actually been provided. Meanwhile, grant military aid was orders of magnitude greater, reaching more than $33 billion in agreements during the same period, virtually all of which was in fact delivered.[2]

The direction of the flow of military aid was largely determined by the alliance structure built up by the United States in the decade after World War II, which was in turn a function of efforts to contain the expansion of communist influence or domination. The earliest and largest of the treaty arrangements were the Rio Treaty of 1947, binding twenty-two Western Hemisphere nations for the purposes of mutual security, and the North Atlantic Treaty of 1949, linking twelve nations, including the United States, with three additional members joining thereafter. More than half of the military assistance provided in this period went to European nations. Conversely, and contrary to the impression apparently held by many, the flow of military assistance to Latin America was always modest, representing well under 2 percent of the total provided worldwide.[3]

Two additional multilateral defense treaties were added in 1951, the Anzus Treaty with Australia and New Zealand, and 1954, the Southeast Asia Treaty involving eight nations directly and three others about which we were later to hear a great deal more—Vietnam, Laos, and Cambodia—as associated "protocol" states. During the decade 1951–1960 four bilateral treaties were added, between the United States and the Philippines, South Korea, Taiwan, and finally Japan, respectively. Largely as a result of military assistance to these diverse allies, the program reached its all-time high in FY1952, when Congress appropriated nearly $6 billion for grant military aid.[4]

From that high point the amounts provided in successive fiscal years for military assistance tailed off in a fairly smooth descending curve as economic assistance and peacetime rebuilding enabled more nations to shoulder increasing shares of the mutual defense burden. Foreign military sales on credit terms became increasingly prominent, to be progressively supplanted in many cases of more recent years by outright cash or commercial credit transactions. The crossover point between grant aid and sales is usually pegged at FY1962, the first full year under the aegis of the Kennedy administration. In that year sales (in the aggregate of both cash and credit) exceeded grants for the first time, a shift that has never since been reversed.[5]

The extensive debate and controversy in the Congress that has surrounded administration requests for grant military assistance in recent years has tended to obscure the declining significance of such aid, in contrast with the vastly greater sums of earlier years. Only in the case of special large grants in the form of waived repayments of loans to Israel has this trend not held steady. Such waivers resulted in some $2.45 billion being provided to Israel between FY1955 and FY1976, nearly all of it concentrated in the post-1973 war period of FY1974–FY1976.[6] Meanwhile, appropriations for military assistance had declined by the time of the last pre-Nixon budgets to just $375 million, an all-time low to that point and only about 6.5 percent of the peak amount in the early 1950s.[7] It should be emphasized here at the outset, however, as it will be again, that these dollar amounts can be taken only as general indications of trends and predispositions, since they lump together arms, other types of military equipment, logistical and construction support, and other categories of assistance and sales. Thus it is important to remember that in discussing such dollar amounts we are far from talking about just transfers of actual arms.

A number of factors account for the crossover to predominance of sales, including the resurgence of foreign economies and the concomitant expression of independent preferences on the part of recipients of arms. Not the least of the influences at work was concern caused in the United States by increasingly unfavorable experience with the balance of trade and balance of payments, the latter aggravated substantially by expenditures for maintaining the overseas military establishment, especially in Europe. The Kennedy administration, having found that, contrary to its campaign assertions, there was no missile gap, turned its attention to problems of conventional force structure. And, along with its interest in developing and fielding modern conventional weapons for United States forces, there was complementary attention to providing, and where possible selling, such weapons to allies abroad.

In the autumn of his first year on the job Secretary of Defense Robert McNamara established an office of International Logistics Negotiations (ILN) to coordinate United States military export sales. One of its sales teams, concentrating on the West German market, was specifically assigned to help offset the costs of maintaining U.S. military forces in that country, which amounted to some $775 million a year. A Senate Foreign Relations Committee staff study of 1967 documented the success of this effort, West Germany having purchased some $3 billion worth of military equipment in the preceding four years.[8] While the worldwide sales promoted by ILN rose manyfold over the levels of earlier years, reflecting what has been characterized as a deliberate policy of increasing arms sales,[9] external factors, without which the sales effort could not have succeeded to such an extent, contributed to this result. These factors included proliferation of states, necessity for independent defense capabilities in nations previously reliant upon colonialist protectors, increasing perceptions of security threats, growing availability of disposable resources, and block obsolescence of inventories of military equipment in the hands of many nations.

There was also a set of attitudes in the community of nations that predisposed sovereign states to seek modern arms. Even in the days of the League of Nations such documents as its 1922 resolution on disarmament and security reflected understanding that reliance on armaments for national security would not be abandoned until some other means of providing such security had been found.[10] But no effective collective security arrangements of global scope evolved, either then or following the Second World War. The trend was such that preparedness, rather than disarmament, came increasingly to be viewed by nations and their leaders as the more viable means of preserving the peace or of prevailing in the event of war.[11] The postwar regional pacts that served to share responsibility for maintaining peace depended preeminently on an adequate supply of arms.

The United Nations sporadically attempted to promote disarmament, but few member nations showed serious interest. As early as 1948 the United Nations Commission for Conventional Armaments had adopted a resolution establishing some general principles, among them the following: "A system of regulation and reduction of armaments and armed forces can only be put into effect in an atmosphere of international confidence and security." Less than two years later the secretary-general, in apparent recognition of the absence of any such guarantees of security or of any prospects for achieving them in the near term, observed that, "while disarmament required an atmosphere of confidence, any progress towards agreement on the regulation of armaments would help reduce tension and thus assist in the adjustment of political disputes."[12] The Disarmament Subcommittee of the UN Disarmament Commission first met in 1954; after three years it suffered a complete breakdown. A successor organization, the Ten-Nation Committee on Disarmament, lasted two years. In the spring of 1960 Secretary-General Hammarskjöld told this body that "the United Nations, like other international organizations, of course reflects only the political realities of the moment." As if in confirmation of his point, the ten-nation talks collapsed.[13] Meanwhile, dozens of new nations were emerging as independent factors in world affairs. Despite their diversity in almost every other respect, they seemed to have one thing in common: they all wanted weapons.

In 1967 the Congress imposed crippling restrictions on foreign military sales after a fierce debate centered on the foreign aid bill. Two amendments sponsored by Congressman Silvio Conte provided that economic assistance must be withheld in amounts equal to that expended by the intended recipient on such sophisticated weapons systems as jet aircraft and missiles and barred the use of military assistance and credit sales revolving funds to help underdeveloped nations finance purchases of sophisticated weapons. A complementary amendment introduced by Senator Stuart Symington required the president to cut off aid to any nation using either U.S. or its own funds for military expenditures that materially interfered with its development. Two sets of hearings conducted by the Senate Foreign Relations Committee in 1967 focused on problems of armament and disarmament and on arms sales to the Near East and South Asian regions, calling into question both the objectives and the methods of U.S. export sales of arms. A staff study done for the committee and published the same year provided additional critical commentary and, in conjunction with the hearings, led to further legislative action the following year.[14]

Thus, by 1968, a major congressional initiative to achieve even greater control over arms sales resulted in enactment of the Foreign Military Sales Act, a bill actually proposed by the Johnson administration as a response to previous congressional criticism in an attempt to modify the stringent restrictions imposed by statute the preceding year. The new law made sweeping changes in the way the United States went about effecting arms transfers to foreign nations. It kept intact the provisions of the Symington amendment of the previous year, and the Conte-Long amendment as well. It barred loans by the Export-Import Bank, previously a common source of funding for credit sales of arms, to less developed countries, providing that such sales be financed solely from appropriations for that purpose. This put a ceiling on credit sales; it also placed regional ceilings on sales to Africa and Latin America, and it charged the secretary of state with overall policy control of arms sales and with ensuring their integration with other United States initiatives in support of foreign policy.[15]

The dialogue on this bill, as has usually been the case in discussions of "arms sales," made no distinction between sales that involved actual arms—such things as weapons and ammunition—and those consisting of other types of military purchases, whether construction, supporting equipment, supplies, or whatever. The single exception involved training of foreign military personnel, which was factored out from the rest of the category.

Administration officials lobbied for the bill on Capitol Hill and before other audiences elsewhere. In testimony before the House Foreign Affairs Committee, Secretary of Defense Clark Clifford argued that "the United States should not— and in the light of free world defense requirements, cannot —avoid continuing to serve as a source of arms supply for those countries whose security is linked to our own."[16] Assistant Secretary of Defense Paul Warnke supported credit assistance to nations wishing to buy arms from the United States, pointing out to the Senate Foreign Relations Committee that such arms help others defend themselves, contribute to free world defense, and make direct U.S. involvement unnecessary and that "if countries friendly to the United States could not buy arms here they would obtain them elsewhere and this meant chiefly from the Soviet Union."[17]

Parallel forces were at work in the closely related area of foreign aid. The program proposed by the administration for FY1969 included $420 million for military assistance, which Secretary of State Dean Rusk had called an austere program.[18] But in 1968 the question of foreign aid generally, and the included security assistance, was very much entangled with issues relating to the war in Vietnam, particularly the cost of that endeavor. Many argued that the United States could not afford foreign aid, given what it was already spending on the war. Others said that it invited involvement in other Vietnam-type conflicts to provide military assistance to client states. While assistance related to the theater of war was at this time in the Defense Department budget rather than the foreign aid bill, the magnitude of such expenditures was an important determinant of the legislative climate.

The Congress had, the previous year, passed the smallest foreign aid bill in the twenty-year history of the program, the result of an alliance of sorts between those who wanted to restrict the flow of arms and those who wanted to reduce

costs.[19] In 1968 another new record low was set, with a billion dollars being slashed from the $2.9 billion administra-
tion request, which was itself by more than half a billion
dollars the smallest amount ever requested.[20] When an earlier proposal for a reduction of $765 million was reported to
the House Republican leader, Congressman Gerald Ford,
"he sent back word that the cut was not enough,"[21] so a
round billion was cut in the authorization process, including
a comparatively modest decrease in the military assistance
component to $375 million. Then, in addition to this third
cut in foreign aid in as many years and the tripling of interest rates on development loans due to congressional action
over the past two years,[22] there were included in the bill
additional restrictive policy amendments. The president was
ordered to reduce and to terminate with "deliberate speed"
military assistance grants to nations that in his judgment
could afford to pay their own way.[23] In the Senate, William
Fulbright was among those opposing the bill, the first time
in history that a chairman of the Foreign Relations Committee had voted against a foreign aid bill.[24]

The Nixon policy on arms transfers had, of course, to take
into account and work within the constraints of these trends,
influences, and attitudes. But it had also to be compatible
with and supportive of the larger foreign policy goals and
initiatives of the new administration. Those in turn were
shaped by and in part responsive to the context of the times,
the international and domestic events and contending interests of greatest current significance.

"To be the prisoner of one's time and place is one of our
human limitations," once wrote Arnold Toynbee. "However,
it is characteristic of our human nature that we rebel
against our human limitations and try to transcend them."[25]
In the same way a government seeks within its times to
achieve goals, devising in the process—if it is to be successful
—means of overcoming whatever obstacles stand in the way.
It is no exaggeration to say that, when the Nixon administration came into office, the difficulties that confronted it were
of heroic proportions. It was a time when international and

domestic events and influences presented unusual challenges to a government that sought to shape and to some extent control the evolving course of events. The strategic environment had been radically altered, perhaps permanently, through attainment of strategic parity by the Soviet Union. The prior administration had been driven from office by elements opposed to further United States involvement in Vietnam. Domestic harmony had been strained severely by riots, arson, bombings, hijackings, and widespread protest demonstrations and movements with origins in various racial, political, and economic grievances. The willingness of the American people to support an activist role in world affairs had been severely eroded by the prolonged, expensive, and ultimately unsuccessful intervention in Southeast Asia. Customary American impatience reinforced this trend, as did, for some, a lessened perception of any external threat requiring a response. And, though it lay in the future, the loss of credibility and eventually of office stemming from the Watergate affair was destined to further compound these obstacles to effective action in world affairs.

The new administration was nevertheless determined to preserve a meaningful role for the United States in world affairs. Late in his life Dag Hammarskjöld had spoken of this as a world with "very great unity of fate and destiny."[26] His words could almost be taken as the scriptural text for the Nixon administration's approach to foreign affairs. This government exhibited at every juncture the conviction that the United States could not turn away from world affairs nor decline to play a part in helping to shape their evolution. Thus, immediately upon taking office, the new administration put into effect an expanded and strengthened National Security Council system. Numerous studies upon which to base policy reassessments in various realms were ordered, and initiatives were put in train that came to fruition as the major milestone achievements of the administration's stewardship. These included the opening to China, the establishment of a degree of détente with the Soviet Union, termination of the war in Vietnam, substantial progress toward peace in the Middle East, and stabilization of areas of critical strategic and economic importance to the United States. Conventional arms transfers played a major and specific role in bringing about many of these successes. In a more general way they were also indispensable to maintaining the capacity to act that made other United States initiatives significant and viable.

At the most general level the process that followed in the wake of World War II, the breakup of colonial empires and proliferation of new and generally small states, had resulted in a United Nations General Assembly dominated by the Third World dialogue. Aspirations for greater economic benefits, and especially for a reduction in the differential in the standard of living between industrialized nations and those less developed, came to permeate virtually every issue. Both the international security system and the international economic system, which had functioned effectively for some two decades, were showing signs of strain and potential disruption. The advance toward strategic nuclear parity by the Soviet Union had made it clear that, at least pending further revolutionary technological breakthrough, there were unaccustomed constraints on the use of force and both political and security implications of a new strategic relationship to be assimilated and adjusted to. The clear and unambiguous guiding image that had shaped American policy, containment of Soviet expansionism and international communism, had progressively eroded in the years following the Sino-Soviet split. While the Soviet invasion of Czechoslovakia the preceding year had dispelled some wishful thinking, other events—and preeminently the prolonged war in Vietnam—had done much to undermine whatever consensus had existed as to the scope and nature of the external threat and America's proper role in responding to it.

In the economic sphere, nations that had been helped to recover by American assistance in the postwar years had developed into impressive competitors for world markets. The United States was experiencing balance of payments problems, domestic interest in at least selective protectionism was resurgent, inflation was a continuing and troublesome presence, and the previous year's balance of trade surplus had been the lowest since 1937. But it was Vietnam that preoccupied the nation. A decade that began with a rash of political assassinations and was punctuated by recurrent racial turmoil, demonstrations, riots, and the burning of sections of American cities neared its end with the war as the most contentious and divisive issue. The political signposts that had read *Watts, Dallas,* and *Selma* now referred to *Danang* and *My Lai.* Soon they would also be marked *Cambodia,* then *Kent State.* The Nixon years were as crowded as any since World War II with turbulent events of dramatic impact and long-term significance. This contextual reality is essential to understanding the policy options available to national leaders and the forces that shaped their perceptions of those options and their likely outcomes.

Six topics stand out on the foreign policy agenda of the Nixon presidency that were not only consistently of highest priority but intertwined in important ways: a start toward ending China's isolation; détente with the Soviet Union and a range of negotiations on such topics as Berlin, the Conference on Security and Cooperation in Europe (CSCE), and Mutual and Balanced Force Reduction negotiations (MBFR); the Strategic Arms Limitation Talks (SALT) as a special and preeminent case of superpower interaction and negotiation; the search for a peace agreement in the Middle East; termination of the war in Vietnam on terms at least minimally acceptable to the United States; and the problems of international economics and reform of the governing mechanisms.

These concerns are, in their totality, the best evidence of an outlook on the part of the new administration that is of central importance to the present study, for each item individually, and particularly the interlocking complex viewed as the central agenda, postulates the continuation of an involved and constructive role for the United States in world affairs. Much of the energy and attention of the administration was to be devoted to efforts to maintain the capacity to act effectively. This meant countering the effects of external events and fending off those who sought to severely curtail American involvement in international affairs. Arms transfers were one of the most important instruments used by President Nixon and his administration in seeking to retain and exercise influence in world affairs.

At the end of the Nixon/Ford period conferees on arms transfers reflected that "many of the current disputes between the Executive and Congress over the size and direction of U.S arms transfers reflect a fundamental debate over the appropriate goals of U.S. foreign policy."[27] The record shows that this controversy was already full-blown by the time of Mr. Nixon's accession to the presidency and that it continued unabated throughout his tenure. In addition to the question of the goals of American foreign policy there was in contention another perhaps even more fundamental question: Who should determine those goals? Thus, throughout this administration there was more or less open warfare between the executive branch on the one hand and those who were determined to restrict its opportunities to involve the United States in overseas affairs on the other.

The continuing contention took many forms. Senator Clifford Case proposed in April 1972 that Congress cut off assistance to Portugal and Bahrain until the executive branch submitted in the form of treaties for Senate approval the base agreements recently concluded with those two nations, contending that "a fundamental constitutional question is at stake."[28] According to one tally compiled in 1970, in addition to the four multilateral and four bilateral mutual defense treaties involving the United States, at least ten defense agreements had been concluded by executive agreement (Denmark 1951, Iceland 1951, Spain 1953, Canada 1958, Liberia 1959, Iran 1959, Turkey 1959, Pakistan 1959, and the Philippines 1959 and 1965), and thirty-four executive branch policy declarations and communiqués had been issued jointly with foreign governments (covering Latin America, Berlin, Iran, India, Jordan, Israel, Thailand, South Vietnam, Taiwan, and the Philippines).[29] Apparently Senator Case, who had been in the Senate since 1955, was slow to recognize, or at least to raise, this "fundamental constitutional question." The Senate majority leader, Mike Mansfield, kept up a personal crusade throughout the period to severely reduce the U.S. military presence in NATO Europe. This too was part of the context in which the president had to operate. A result of the pressure was that he had to place stress in his proposed reorganization of the foreign assistance program, and particularly the security assistance aspect of it, upon the potential for reducing the involvement of U.S. forces abroad and upon the fact that forces in Vietnam, Korea, and elsewhere were already being drawn down as a result of the assistance being provided. The issue of reducing troop strength in Europe became symbolic of the contest between those who wanted to diminish U.S. involvement overseas and the administration, which opposed unilateral reductions without any diplomatic reciprocity.[30]

Commentators remarked upon the way in which the Congress was beginning to question basic assumptions about U.S. overseas commitments and involvements and to assert itself more in the realm of defense and foreign policy. One journal reported an atmosphere of profound change, with the key question being whether President Nixon could develop sufficient political support to succeed in his attempts to maintain an active U.S. part in international affairs while scaling down some dimensions of that role so as to satisfy domestic critics.[31]

There was, to be sure, a great deal that was ambivalent in the attitudes of many. While the role of guarantor of international order was no longer uniformly popular, in selected cases strong constituencies demanded that the United States take action to influence foreign developments, such as the nature of the regime in Greece, the policies toward blacks of the South African government, or the prospects for peace in the Middle East. Seldom did these advocates concede any inconsistency between opposing provision of the means of intervention in the general case and demanding results—which depended upon the application of such means—in selected specific instances.

The role of the press during this period could itself be the subject of volumes, as indeed it has, in certain respects, already been. The Nixon-press relationship had never been cordial, and the famous "last press conference" statement by Mr. Nixon following his defeat in the 1962 California gubernatorial election, observing that the press would not have Nixon to "kick around any more," had long since passed into the lexicon of famous American catch phrases. No doubt Lyndon Johnson's monumental struggles with the press did much to set the scene for continuation of an adversarial relationship when Mr. Nixon came into office. But beyond that the Nixon inner circle was widely believed to conceive of the press as the "enemy."

These predispositions were strongly reinforced by the all-pervasive influence of the various information media in reporting two key long-running stories, Vietnam and Watergate. In the case of the war, television clearly played the dominant role, whereas major newspapers led the way in reporting on Watergate. Both were reinforced by influential journals and by one another, as in the case of the major impact on the Vietnam story resulting from newspaper publication of the "Pentagon papers." No definitive judgments are possible at this juncture with respect to the performance of the press taken as a whole during the Nixon years, although some responsible studies have documented systematic bias in reporting on the war.[32] But the press was undeniably a major factor in determining both the focus and the tone of domestic dialogue during the period.

It may be objected that there is nothing new in this, and of course there is some truth to that, with the differences, if there were any, being more of degree than kind. But there was also the feeling, which persists in retrospect, that the role of the press had somehow significantly changed. Political satirist Garry Trudeau expressed it through the words of his *Doonesbury* cartoon character, rock music star Jimmy Thudpucker, in an "interview" for *Rolling Stone* magazine: "But I'll tell you, in the long run, the fallout from [Carl Bernstein and Bob Woodward's] Watergate reporting could have a very negative effect on their craft. What has happened is that for the first time in the history of the Republic, reporters are starting to become as important as their stories."[33]

The war in Vietnam served as focus for much of the activity in Congress aimed at restricting American involvement overseas and the authority of the executive branch to enter into commitments without prior approval of the legislature. In addition to the skirmishes over foreign aid and the defense budget generally, increasingly there were attempts to constrain the war effort. The Cooper-Church amendment, passed in the wake of the incursion into Cambodia, restricted further such activity in June 1970. Ten days later the Senate repealed the Tonkin Gulf Resolution, which had been passed in 1964. Early the next year measures were enacted barring use of U.S. ground troops in Laos and Cambodia. The next year aid to Cambodia was severely restricted, and subsequently the entire foreign aid bill was rejected. That rejection was accompanied by denial, in 1971, of the proposed military assistance program, leading the *Christian Science Monitor* to observe that "the future of the Nixon doctrine is now a wide-open question," since that program was to have provided the means to substitute for direct involvement of U.S. troops and equipment.[34] The year after that, following the cease-fire agreement, funds were cut off for any U.S. military operations in Indochina, including bombing in Cambodia, without specific congressional authorization. Later in the year the War Powers Act, limiting the president's authority to commit armed forces to foreign hostilities, was passed after Mr. Nixon's veto was overridden.

In a further step toward limiting presidential powers the Congress in September 1976 passed the National Emergencies Act, which revoked a range of standby powers that had long been available to the chief executive in various crises. In relating how "very satisfying" he found this revocatory legislation when it came into effect in September 1978, Senator Charles Mathias suggested that we had thereby averted the possibility of an authoritarian takeover of the U.S. government of the kind perpetrated by Hitler in Germany. Reporting to his constituents on the work he and Senator Church had done as cochairmen of the committee that initiated the bill, Senator Mathias concluded that thanks to their efforts the United States had "been returned to the rule of law for the first time in more than 40 years."[35]

Some more recent analyses have detected a modest reversal of the sentiment for substantial retrenchment in the American role that peaked in 1973–1974, when the Vietnam War, racial strife and its aftermath, and the Watergate affair conjoined to markedly diminish America's confidence and expectations.[36] But at that low point a remarkable calendar of significant events transpired, from negotiation of a cease-fire in the Middle East and cessation of the oil embargo to the end of United States involvement in Vietnam, and from summit conferences with the Soviets to CSCE and MBFR negotiations in Europe, and including reestablishment of diplomatic relations with Egypt and Syria. Not incidentally, the United States also acquired one new president and two new vice-presidents (successively) during the period. In fact, October 1973 may have been the single most eventful month in U.S. history: While war was breaking out in the Middle East, then being halted by a cease-fire that came undone, to be succeeded by a second cease-fire two days later, with associated arms shipments and potential confrontations between the superpowers and an oil embargo imposed by Arab producers, MBFR talks opened in Vienna; Secretary Kissinger visited Moscow to negotiate with the Soviets on plans for a Middle East settlement and avoidance of escalation, and Golda Meir visited Washington, where one vice-president had just resigned and another had been appointed; the House of Representatives began an impeachment inquiry; the attorney general and his deputy resigned rather than carry out the president's instructions to dismiss the Watergate special prosecutor, who was then sent packing by the solicitor general; and Congress passed the War Powers Act, which the president then vetoed (his veto would be overridden, but not until the *next* month). In the midst of it all Henry Kissinger and Le Duc Tho were named as corecipi-

ents of the Nobel Peace Prize. By any standards it was quite
a month.

Yet another aspect of the policy context that had something to do with how effectively and to what purposes various instruments of policy, including arms transfers, could be
used was the change in relations among individual states
and groups of states. Militancy on the part of Third World
nations in particular was a factor of growing importance.
Manifestations of increasing resistance to major power influence made arms transfers, which recipient nations strongly
desired and considered of crucial importance to their interests, more prominent as a means of maintaining access and
influence.

Such developments as observed limitations on the utility
of force, at least for a self-constrained nation such as the
United States; newly demonstrated vulnerabilities of dependence on foreign energy resources; evolution of aspects of the
Western alliance into economic competitiveness and to a
lesser extent political diversity of outlook; proliferation of
nation states and increasing attention demanded by North-
South issues, however much their real significance depended
in many cases upon United States forbearance—all these
formed parts of what could be termed a general transformation of the international system. But the central aspect of
that transformation, and in many ways the key determinant
of the international context of the Nixon years, was the
military and political balance between the United States
and the Soviet Union.

"[Soviet] missile power will be the shield from behind
which they will slowly, but surely advance through Sputnik
diplomacy, limited brush-fire wars, indirect non-overt aggression, intimidation and subversion, internal revolution,
increased prestige or influence, and the vicious blackmail of
our allies. The periphery of the Free World will slowly be
nibbled away. . . . Each such Soviet move will weaken the
West; but none will seem sufficiently significant by itself to
justify our initiating a nuclear war which might destroy
us."[37] So warned John F. Kennedy in 1960, and there were
many who believed him and many who were, at least for a

time, willing to try to do something about it. But over the course of a decade and a half the price began to seem too high to some, others began to feel a sense of impotence in terms of American ability to affect the course of world events, and others at least professed to see a reduction in the threat, implying that the need for an activist stance no longer existed.

By the point at which Mr. Nixon came into office the nuclear weapons balance between the United States and the Soviet Union had reached near equality. This was, as a Defense Department official of the period later pointed out, "one important reason why the Nixon Administration began to talk about a 'new balance of power in the world.' "[38] Two additional significant developments were underway: contrasting trends in defense expenditures between the United States and the Soviet Union, and major reordering of budgetary priorities in the United States. While Soviet military spending, adjusted for inflation, was reported by the Arms Control and Disarmament Agency to have risen from the equivalent of $80 billion in 1967 to $121 billion in 1976, adjusted U.S. military spending over the same period fell from $120 billion to $87 billion.[39] This occurred despite very large U.S. expenditures for the war in Vietnam, which did little to promote residual military capability and in fact had a devastating effect on worldwide military readiness. American military outlays as a percentage of gross national product also declined steadily from the 1968 Vietnam War peak of 9.1 percent, reaching just over 5 percent by the end of 1976. By way of contrast, U.S. military spending at the height of the Korean War had peaked at some 14 percent of GNP.[40] The significance of this decline is even greater in terms of actual military capability being purchased, since the end of the military draft in 1973 and the shift to volunteer armed forces were accompanied by very large increases in personnel costs, to the extent that such costs came to absorb well over half of the total defense budget, severely squeezing the investment accounts.

Considering the period bounded by Mr. Kennedy's warning on the one end and the passing of the Nixon/Ford administrations on the other, a dramatic change took place in shares of the federal budget allocated for various purposes. While total federal outlays increased in constant dollars by 117 percent over the years 1960–1977, defense spending rose 5 percent and nondefense spending 225 percent. Those aspects of the budget categorized as social spending, such as education, health care, income security, housing, and employment programs, rose 322 percent. In addition to the shifts in priorities reflected in these disparate growth patterns, the extent of the major reordering is clear from examining respective shares of the federal budget. In 1960 defense and nondefense expenditures accounted for roughly equal portions of the total budget (49 percent for defense versus 51 percent for nondefense). By 1977 defense expenditures amounted to only 23.7 percent of the budget, whereas nondefense comprised 76.3 percent. This shift was paced by increased spending for social programs, which had come to take a much larger share (55.9 percent in 1977 versus 28.7 percent in 1960) of a federal budget that had, over the period, more than doubled in real terms.[41]

These steady changes in relative emphasis on defense spending took place concurrently with more or less continuous negotiations with the Soviet Union on a range of issues, some resulting in agreements of major significance (SALT I, CSCE, the hot line, limited test bans, nuclear nonproliferation, seabed and outer space nuclear prohibitions), others continuing for extended periods without agreement being reached (SALT II, MBFR, trade agreements). At a Moscow summit meeting in 1972 the United States and the Soviet Union agreed to a "Statement of Basic Principles of Relations." Later described by Marshall Shulman as "a codification of the ground rules for a competition for political influence around the world in which the use of strategic weapons would be subordinated,"[42] it left plenty of room for continued competition between the superpowers. This competition would increasingly rely on the use of surrogates, who would in turn have need of the military wherewithal to perform effectively. The importance of arms transfers to maintaining influence abroad received reinforcement and increased emphasis as a result.

In this context the Nixon administration sought to shape and implement a coherent foreign policy. The major nations, including the United States, had more power than ever before, but there were also more inhibitions for some on the use of it, with the ultimate concern that any conflict involving the superpowers might escalate into nuclear war. This led to the search for surrogate means of bringing power to bear. Among the approaches found useful in various situations by one or both superpowers were provision of foreign aid, the use of proxies, subversion, and the creation of alliance structures.

During American involvement in Vietnam the strategic balance had evolved into what was perceived as a stalemate, after an extended period of American dominance; likewise, early in the period the conventional force balance was essentially a standoff. It was in the realm of what might be called "subconventional" conflict, of insurgency and guerrilla warfare, that the United States at that time saw itself challenged, with Khrushchev's vows of support for "wars of national liberation" reinforcing the view. A decade later the experiment in confronting such a threat in Vietnam had proven unsuccessful, and one result had been severely diminished support for the military establishment and far less enthusiasm for an activist American role in world affairs. This fact, reflected in defense budgets and reordered national priorities, meant that America's allies needed to be more capable of shifting for themselves, a reality that was both recognized and reinforced by the Nixon Doctrine. The imperative was, therefore, that these allies be more extensively, and expensively, armed.

It is for these reasons that the development and functioning of the Nixon administration's arms transfer policies must be viewed in the context of the international and domestic forces at work on them and the available decisions and options that resulted. The administration was faced with preserving what it could of freedom of action and the ability to influence world events. If direct U.S. involvement was not to be supported, if the conscription of U.S. forces was no longer politically viable, and if foreign assistance bills became slimmer and slimmer and required longer and longer wrangles to get them enacted, then the sale of arms to those it was desired to support became one of the few remaining sources of influence in many places in the world. That this was an imperfect instrument, with results not always calculable in advance, was less important in the circumstances than that it offered some prospect of beneficial impact. As Henry Kissinger had pointed out long before, in foreign policy "only the risks are certain; the opportunities are conjectural."[43]

Some of the administration's perspective on the national mood that had confronted it came out in a revealing conversation among Dr. Kissinger, Senator Moynihan, and Ben Wattenberg something over a year after the Ford presidency had ended. The topic was whether there was a crisis of spirit in the West, something that both Kissinger and Moynihan had addressed from time to time. Mr. Moynihan pointed out that with Vietnam we had an enterprise begun by a very confident and powerful bipartisan foreign policy elite that had been in command for a quarter of a century and had been enormously successful. When Vietnam failed, he said, something even more significant occurred. "The social base on which that elite rested turned against its own policies." Dr. Kissinger agreed, adding that among that group and also to some extent among the intellectuals of the same generation there developed "not just a disappointment with a lost war but almost a *desire* to lose. . . . it turns into a self-hatred." The challenge to the administration was to do what could be done given the circumstances of the moment. "In that period," Kissinger went on, "there would have been no support for a crusading policy as you have suggested; our worry was that the assault on our foreign policy would collapse *all* our commitments." The reality is, he said, that "leaders have to conduct the policies that circumstances make possible for them."[44]

2. The Nixon Foreign Policy

Analysis of the record reveals certain fundamental precepts embodied in the foreign policy of the Nixon administration:

First, and controlling, was the belief that interests depended on maintenance of a stable world order that could accommodate basic changes evolving in international relationships.

Second was the conviction that, to further its interests, the United States must continue to play a meaningful—and therefore activist—role in world affairs.

Third was the premise that effective foreign policy must be guided by an overall concept, which was often referred to by this administration as building "a durable structure for peace."

Fourth was the idea that major aspects of foreign policy are necessarily interrelated, and thus can be dealt with effectively only if these connections, or linkages, are recognized and taken into account.

Fifth was the determination to meet the necessity for communicating the premises and aspirations of the foreign policy to the American people so as to build understanding and support.

These precepts were operationalized in the development and implementation of a foreign policy that emphasized in its grand agenda these key topics: China, the USSR, SALT, the Middle East, Vietnam, and international economics. The tasks of policy formulation had at their heart the resolution of an essentially philosophical problem, as the preceding

discussion of context has sought to illustrate. Lincoln Bloomfield had formulated it concisely in laying out a suggested
strategy for the 1970s, which was published the year Mr.
Nixon took office: "The bewildering riptides of contemporary
U.S. policy arise largely from the clash between traditional
American political and moral sentiments and the realities of
exercising power and asserting responsibility."[1] Those
clashes, and attempts to deal with them and to act despite
their impact, were to persist throughout the administration's tenure and were to form much of the material of the
continuing debate over arms transfers.

It has been suggested that the way in which the United
States views its role in the international system should be
the dominant factor in the arms transfer planning process.[2]
That idea appears to be at the heart of much of the contention over arms transfers as utilized by the Nixon administration, for those opposing the policy seemed often to be
operating from a different set of premises as to what the
United States could and should do to try to influence international affairs. While the interplay of ideas is an important,
and built-in, part of the functioning of the American system
of government, that process can work only when there is at
least a modicum of prior agreement as to where the country
is trying to go, with the bulk of the discussion devoted to how
it should go about trying to get there. When the goals themselves are very much at issue, the result is disabling to policy
in many instances. This is an important factor in assessing
the Nixon arms transfer policy, and one that makes what
that policy achieved all the more significant, given the fundamental obstacles that had to be overcome in the course of
formulating and implementing it.

The broad outlines of overall foreign policy were set forth
initially in the first inaugural address and importantly aug-
mented by the first suggestion of the Nixon Doctrine at
Guam and later elaborations of the specifics and implications of that doctrine. But perhaps most important as a vehi-
cle for setting forth publicly both the policy goals and the
supporting rationale was the series of four annual reports
to the Congress on foreign policy that Mr. Nixon introduced

in 1970. Billed by some as "State of the World" addresses, they illustrated and explained the policy implications of the precepts we have identified and showed how they applied to the elements of the grand agenda.

Mr. Nixon had anticipated the difficulties of implementing an activist foreign policy. As he observed in *Foreign Affairs* in October 1967, the war in Vietnam had "imposed severe strains on the United States, not only militarily and economically but socially and politically as well," thus raising doubts on the part of friendly countries as to America's willingness to intervene on their behalf should their security be threatened in the future.[3] The clear implication was that nations that had relied on the United States to underpin their security had better seek to become more self-reliant. As far as American foreign policy in this context was concerned, the obvious imperative was to devise means of helping friends and allies become better able to defend both themselves and those interests they and the United States had in common.

In a foreign policy address of 3 November 1969 President Nixon reiterated what had evolved as the key principles of his doctrine:

> First, the United States will keep all of its treaty commitments.
> Second, we shall provide a shield if a nuclear power threatens the freedom of a nation allied with us or of a nation whose survival we consider vital to our security.
> Third, in cases involving other types of aggression, we shall furnish military and economic assistance when requested in accordance with our treaty commitments. But we shall look to the nation directly threatened to assume the primary responsibility of providing the manpower for its defense.[4]

This was the famous speech in which Mr. Nixon appealed for support to the "silent majority" of the American people during the course of his pursuit of peace in Vietnam. While the policy described as the Nixon Doctrine was specified as applying to Asia, which was understandable given the necessary preoccupation of the moment with the war in that theater, its potentially broader application was obvious. Within an international context that itself strongly encouraged sharp increases, albeit on a highly selective basis, in the volume, value, and sophistication of arms transferred, the utility of arms transfers to accomplishing the administration's goals in foreign policy and the restrictions placed on alternative means served to further extend that contextual tendency.

Upon conclusion of the first year in office, Mr. Nixon presented to the Congress his first extensive report, *U.S. Foreign Policy for the 1970's,* subtitled *A New Strategy for Peace.* "We could see," the president said, "that the whole pattern of international politics was changing. Our challenge was to understand that change, to define America's goals for the next period, and to set in motion policies to achieve them. For all Americans must understand that because of its strength, its history and its concern for human dignity, this nation occupies a special place in the world. Peace and progress are impossible without a major American role."[5]

Outlining the changes on the world scene since the aftermath of World War II, the president observed that "others now have the ability and responsibility to deal with local disputes which once might have required our intervention."[6] With regard to the Nixon Doctrine, then, he cited as its central thesis "that the United States will participate in the defense and development of allies and friends, but that America cannot—and will not—conceive *all* the plans, design *all* the programs, execute *all* the decisions and undertake *all* the defense of the free nations of the world. We will help where it makes a real difference and is considered in our interest."[7] Thus, he said, "the only real issue before us is how we can be most effective in meeting our responsibilities, protecting our interests, and thereby rebuilding peace."[8]

Each of the four volumes in this series was to bear a subtitle reflecting the concern for peace: a new strategy for peace in the case of this initial effort, followed in succeeding years by rebuilding for peace, the emerging structure of peace, and shaping a durable peace. The implication of evolution and progress when those subtitles are taken sequentially is not accidental, no doubt, but there is substantial evidence for making such a case. One recalls Henry Kissinger's point in his early *Foreign Affairs* article to the effect that "peace . . . cannot be aimed at directly; it is the expression of certain conditions and power relationships. It is to these relationships—not to peace—that diplomacy must address itself."[9] That outlook is reflected in the detailed concern with relationships—global, regional, and bilateral— that permeated each segment of this serial essay on American foreign policy goals for the decade ahead.

A crucial point, one which was at the heart of much of the controversy over both policy and operational approaches over the next five years, was also here laid down with clarity and simplicity: "We are not involved in the world because we have commitments; we have commitments because we are involved."[10] From the issue of American troops stationed in Europe and Korea to the question of foreign assistance, it appeared that fundamental disagreement over this proposition was a central factor dividing the president and his opponents.

Given the broad outlines of foreign policy as articulated by the president and his senior associates, the translation into actions was not long in coming. In fact, as even a bitter critic conceded, "by the middle of 1969, halfway through his first year as president, Nixon had engaged the United States in a series of major foreign policy initiatives around the world. American policy was more active, searching, intellectually aggressive, and diversified than at any time since the end of World War II."[11] Supporting numerous international contacts and a broad range of policy initiatives was a revitalized, expanded, and strengthened National Security Council staff under Dr. Kissinger. Typical studies in the field of arms transfer policy included considerations of arms packages for

Jordan, military assistance to Israel, the arms embargo of South Africa, technology transfers associated with military exports, and military supply policy toward various states. Similar studies ranged across the full spectrum of foreign policy issues. Critics were later to charge that the administration had no comprehensive arms transfer policy and that decisions on arms sales in particular did not receive full or careful consideration or were not taken in full appreciation of the larger foreign policy issues involved. In this they were wrong. While senior officials might, as was their prerogative, on occasion override the staff recommendations they received, the process of analysis that underlay major policy decisions, whether related to arms transfers or otherwise, is on the record probably the most systematic and orderly we have known.

Relying upon this process, the new administration had undertaken during its first year initiatives toward peace in the Middle East, extensive public and private efforts to resolve the war in Vietnam, preparations for an opening to China, a scaling back of the U.S. defense effort overall, and strategic arms negotiations with the Soviets. Again one of Mr. Nixon's harshest critics acknowledged the president's "immensely ambitious foreign-policy objectives," conceding that he "presented the United States and the world with a spectacular, innovative, and controversial foreign-policy performance." The result was that "during the Nixon years the United States went a long distance to adjust its position to the global realities of the 1970s."[12]

Early on the president related security assistance to the success of the Nixon Doctrine, revealing that he intended to propose a revised International Security Assistance Program "to provide effective support for the Nixon Doctrine."[13] At this point plans had been laid for an expanded military assistance program designed for that purpose, with expenditures to average $600 million annually for the next four or five years.[14] But opponents in the Congress were, as it turned out, going to be able to effect very large cuts to levels far below these projections. Leading the fight were such Senate figures as Majority Leader Mansfield, who responded to the incursion into Cambodia in May 1970 by vowing to vote against all foreign aid in the future.[15]

There had of course also continued to be much discussion of the meaning and impact of the Nixon Doctrine, and the president returned to a lengthy articulation of his thoughts on it in the second foreign policy report, which was sent to the Congress in February 1971. Among his main points were these:

> It will take many years to shape the new American role. The transition from the past is under way but far from completed. During this period the task of maintaining a balance abroad and at home will test the capacity of American leadership and the understanding of the American people. . . .
>
> The Nixon Doctrine . . . should not be thought of primarily as the sharing of burdens or the lightening of our load. It has a more positive meaning for other nations and for ourselves. In effect we are encouraging countries to participate fully in the creation of plans and the designing of programs. They must define the nature of their own security and determine the path of their own progress. . . .
>
> We recognize that the Doctrine, like any philosophic attitude, is not a detailed design. In this case ambiguity is increased since it is given full meaning through a process that involves other countries. When other nations ask how the Doctrine applies to them in technical detail, the question itself recalls the pattern of the previous period when America generally provided technical descriptions. The response to the question, to be meaningful, partly depends on them, for the Doctrine's full elaboration requires their participation. To attempt to define the new diplomacy completely by ourselves would repeat the now presumptuous instinct of the previous era and violate the very spirit of our new approach.

In referring to the new program for security assistance that he planned to present, the president said this:

> It will place increasing emphasis on fostering self-reliance of those with whom we are engaged in a cooperative effort. We will encourage them, and give them the technical assistance needed, to determine their own requirements and to make the hard decisions on resource allocation which a meaningful security posture demands.[16]

The president was then at the beginning of what he would later call the "watershed" year, and he provided an insight into the sense of continuity and stewardship he felt: "In the field of national security, each Presidency is a link in a chain. Each Administration inherits the force in being. The long-range investments made by earlier Administrations define the ability to change that force in the near term."[17] This consciousness and sense of responsibility quite obviously shaped the administration's determination to preserve, through a period in the American experience when war-weariness and confusion of purpose had temporarily eroded the national will to meet the responsibilities of world leadership, the means and relationships through which that leadership could once again be exerted, to find surrogate means and to shape new relationships that in the meantime could enable this administration to continue to exert such leadership to the maximum extent possible, even if at best that were to be at the sufferance, as it were, of the American people.

The dual commitment was to strength and negotiation. On the larger scene, the approach found its meaning in a thoroughgoing review of national security goals and the strategy and force structure implied for achieving them. The strategic nuclear force goal was defined as "sufficiency" rather than superiority, essential equivalence, or parity—terms used by other administrations to signify their approach to the strategic balance. The desired conventional force capability was specified as that sufficient for a "one-and-a-half-war" contingency, scaled back from a two-and-a-half-war posture, which had previously been sought. There was not, it should be noted, any wholesale reduction in forces implied by this decision. Rather, it was a scaling down of the stated force goals to render them more in accordance with the realities of existing United States capabilities.

It was believed by the administration that the strategic balance was such that greater importance would attach to conventional forces and that the United States was liable to military challenges in the conventional realm, as were its allies, particularly in the wake of the defeat in Vietnam. Admiral Thomas Moorer testified to this effect in supporting the administration's proposed military assistance program in July 1973. The implication of changing realities as he saw it was that we were forced to place greater reliance on allies and must therefore ensure that those allies had the necessary quantities and quality of equipment to be able to meet the opposition.[18]

In many ways, this was at the heart of it. Those opposed to an activist role for the United States in world affairs tried to whittle down U.S. forces, cause them to be brought back from overseas, reduce military assistance to allied nations, and finally even restrict the selling of military goods abroad. In the face of this wave of disengagement sentiment, the Nixon administration conducted what can only be viewed as a brilliant rearguard action. Arms transfers played a major role. Mr. Nixon was quite candid when he reminded the nation in his second foreign policy report that he had "repeatedly emphasized that the Nixon Doctrine is a philosophy of invigorated partnership, not a synonym for American withdrawal."[19] While this assertion was discounted by some, it was quite clear to others, some of whom were in opposition because they saw in the doctrine a means of evading their efforts to restrict American influence abroad and thereby minimize American involvement. One such critic who perceived the full implication was Earl Ravenal, who laid it out succinctly for the benefit of the Senate Foreign Relations Committee in urging a move toward zero security assistance:

This administration has never departed from its characteristic rationale for military assistance. Military assistance is [as the president has been saying from the beginning] the inevitable concomitant of the Nixon Doctrine. . . .

The Nixon doctrine is not a program for American disengagement. . . . Rather, the Nixon doctrine is a program for force substitution. It attempts to support all our commitments—and even add a few—while yielding to the stringencies of inflation and rising manpower costs by trimming the force structure. The contradiction in this approach has not escaped notice. The only solution for the administration, logically and practically, is vastly increased military assistance—that is, total arms transfers by any means, overt or covert, in or out of the U.S. defense budget, by new or used arms, gifts or loans, government or commercial sales, by leasing or simply abandoning ships, real estate, and supplies.[20]

Meanwhile, the administration was busy changing the priorities reflected in the federal budget. In FY1969 defense expenditures had accounted for 44 percent of outlays while human resources programs got 34 percent. Mr. Nixon proposed for FY1971 a budget allocating 37 percent of outlays to defense and 41 percent to human resources, a major shift in a short time, and one made even more dramatic by the fact of continuing hostilities in Southeast Asia.[21] The president pointed out that spending for national defense would at that rate take a smaller proportion of the budget than at any time since 1950. Nevertheless, as he told the Congress in his first foreign policy address, "America cannot live in isolation if it expects to live in peace. We have no intention of withdrawing from the world."[22] Clearly, then, other methods of influencing developments would have to be found to compensate for the scaling down of the defense establishment. Secretary of State William Rogers reinforced the point: "This is the keystone of the Nixon Doctrine. We are pledged to reduce the United States military presence overseas. We are also pledged to remain a world power and to honor our commitments. The link that joins these two elements is the principle of self-help."[23]

Other administration officials testified to the same effect, leaving one to wonder how later critics could profess surprise over escalating arms transfers. Deputy Secretary of Defense David Packard, for example, told the House Foreign Affairs Committee that "a higher level of U.S. military aid to underdeveloped countries will be required under the Nixon doctrine of national security" and that arms sales to underdeveloped countries would continue to raise. If the United States were to reduce its involvement overseas and cut back on expenditures, he said, "we must continue, if requested, to give or sell [allied or friendly nations] the tools they need for this bigger job we are asking them to assume."[24] The administration therefore set about devising an arms transfer policy at once supportive of its foreign policy and attuned to the changed domestic and foreign contexts.

3. Arms Transfer Policy and Mechanisms

The purposes for which arms transfers have been made are numerous and diverse. At the most general level, as we have observed, they may constitute an instrument of policy that can substitute for more direct involvement on the part of a nation that is inhibited from taking a direct hand. They may be employed for reasons that are predominantly political, military, or economic in nature. Typically, however, particularly where the use of arms transfers by the United States during the Nixon era was concerned, a combination of these factors led to the decision to effect a given arms transfer, with one or another of the considerations having greater weight in various situations.

Probably the typical case for the United States has been that in which arms have been transferred with the intention of strengthening the military capability of the recipient state, thereby furthering the collective defense capabilities of an alliance in which both parties share. The early transfers to Turkey and Greece during the period in which the Truman Doctrine was proclaimed following World War II were of such a nature, as were—and are—the transfers to European nations allied with the United States in NATO. In the contemporary period a large portion of the grant military assistance provided to nations around the world has been for the primary purpose of enhancing their military capabilities, with emphasis on those nations designated as forward defense countries in the period when containment of expansionist communist states was the coalescing factor in security planning worldwide.[1]

A tabulation of the objectives of arms transfers suggested by administration spokesmen and other interested commentators from time to time would have to include at least the following imperatives, which are probably not an exhaustive list:

- Enhance attainment of U.S. global security objectives.
- Reinforce regional stability.
- Preempt or provide alternatives to Soviet arms supplies.
- Gain access to recipient states and their governments.
- Help recipients resist subversion and maintain internal control.
- Obtain base rights and other privileges in the recipient country.
- Improve the military capability of the recipient.
- Demonstrate political or military commitment to the recipient.
- Obtain favorable trade or raw material concessions.
- Offset the cost of stationing U.S. troops abroad.
- Keep U.S. production lines open and provide jobs in the United States.
- Spread development costs and reduce unit costs of military items.
- Recycle petrodollars.
- Ensure access to petroleum and other critical resources.
- Earn foreign exchange and help with balance of payments.
- Assist in deterring hostile forces.
- Spread the burden of common defense.
- Promote collective security.
- Encourage support of U.S. foreign policy objectives.
- Substitute for direct involvement of U.S. forces.
- Increase standardization of equipment used by the United States and its allies.
- Increase the prestige and viability of the recipient government.

- Discourage proliferation of nuclear weapons.
- Acquire political leverage or influence.
- Demonstrate political will and reliability as an ally.
- Avoid adverse effects of refusing to provide requested arms.

Secretary of Defense Melvin Laird laid down the criteria of national interest that were to be applied in reaching decisions on both military assistance and foreign military sales in the Nixon administration. An early memorandum to elements of the Department of Defense included these key points:

> No sale shall be approved unless it is consistent with the foreign policy interests of the United States.
>
> Sales of military equipment to friendly countries will be authorized only after careful consideration is given to the potential impact on social and economic development and on arms races.
>
> Subject to questions of security, foreign policy, and availability, we will sell military equipment to our friends and allies where and when and to the extent needed; we will urge on no friend or ally the purchase of U.S. equipment when it is not needed, when there are better alternatives or when there are higher priority social and economic claims against limited funds.[2]

The revised Department of Defense directive on arms exports issued 10 March 1970 contained an important policy change: "It is the policy of the Department of Defense consistent with over-all national policy and the protection of security interests of the United States to permit the export to friendly nations of munitions articles and services and related technical data including manufacturing license and technical assistance agreements."[3] The earlier version of the regulation, which it replaced and which dated from 1954, had specified that it was the policy to *promote and encourage* such exports, not just to permit them as was now to be the case.

The detailed set of guidelines developed for consideration of arms transfer proposals, like the diverse objectives arms transfers may on occasion be designed to serve, contain high potential for coming into conflict with one another in many cases. In practice, therefore, assessments of the relative advantages and disadvantages must be made in each individual case, with judgmental decisions made as to the weights to be assigned to the relevant factors. Inevitably the results are not always as hoped for or planned, but that inescapable reality is common to many instruments of policy.

While arms transfers have many uses, they are also like many other instruments of policy in having inherent limitations. States cannot be induced to act contrary to their perceived interests on matters they consider of central concern, either by arms transfers or by other means. There are also in certain cases disadvantages in arms transfer relationships. The case of Turkey, for example, will illustrate how political elements can seek to limit arms transfers in an effort to attain more ambitious goals, the limited leverage provided by arms transfers in some cases, the kind of reverse leverage that may be acquired by an arms recipient that has provided some quid pro quo, and the necessity for having authority to carry out a coherent policy if an administration is to succeed.

It is convenient to talk of arms transfers, when what is really being addressed is the transfer of military materiel and services, a more inclusive category and one that in many cases has little to do with anything we might normally consider arms. In the last year of the Ford presidency, for example, while export sales of military goods and services totaled $8.7 billion, only 36 percent of that amount actually consisted of weapons systems, ammunition, and the like. The remainder was devoted to things such as spare parts, supporting equipment, and services, the latter category including construction.[4] If one were to attempt to use rising dollar amounts of military transfers to gauge changes in military capability, it would be misleading to take the gross amount, for much of it would have nothing to do with acquisitions that improve or expand military capabilities. In fact, even knowing the specific systems involved in the transfers is not always enough to support any definitive judgments as to military capability. Whether the weapons acquired are additions to the inventory or replacements for combat losses or obsolete equipment makes a critical difference, for example. Even more difficult to judge is the likely combat effectiveness of the systems in the hands of recipients, judgments that must consider the proficiency of the recipients in operating and maintaining the weaponry, the adequacy of support facilities, and even the appropriateness of the equipment acquired to the security missions important to that country. Dollar amounts of transfers are poor gauges of any of these variables.

Figuring the impact of actual arms transfers depends upon an appreciation of regional military balances, which in turn requires the ability to make a distinction between those transactions that enhance military capability (additional aircraft, for example, or more effective antitank missiles) and those that merely preserve an existing capability (maintenance training or replacement parts) or make possible the substitution of indigenous personnel for those provided by the supplier nation. The support element in the transfer of military equipment and services has become increasingly important as transfers to developing countries have increased. Whereas NATO countries enjoy well-developed infrastructures to support their military forces and can thus concentrate their military imports on military hardware, developing countries need much more in the way of construction, supporting facilities, and training to go along with the weapons systems they acquire. As the export trade shifts to these nations, the already substantial portion of the dollar totals represented by other than arms may be expected to increase.[5]

The Department of State has responsibility for making policy with respect to arms transfers, subject to approval of the president. The Department of Defense has subsidiary responsibility for executing the policy decided upon. Elaborate mechanisms have evolved over the years for bringing about coordination of views when decisions are being considered.

Some observers have held that the Defense Department has played too influential a role, in effect making policy rather than just participating in the process. Paul Warnke, the key Defense official in the field as assistant secretary for international security affairs in the Johnson administration, has denied this. "Military sales is an effective and relatively inexpensive implement of United States national security policy," he told *Armed Forces Managment* in 1968. "It is also necessary to understand that these sales are not based on unilateral decisions by the Department of Defense. Rather, they are a key aspect of foreign policy, made only through close consultation and evaluation with the State Department and other interested agencies."[6] Students of the Nixon years know that, if this were the case before the advent of Henry Kissinger, it is very likely to have been even more so once he took responsibility for all major decisions affecting foreign policy.

The Foreign Military Sales Act was passed by Congress in 1968, and it conferred upon the president certain responsibilities. Mr. Nixon subsequently issued an executive order delegating some of these to others; what he retained was final-approval authority on major policy issues. The secretary of state remained responsible for general supervision of policy. The secretary of defense was assigned general operational duties. Consultation on arms transfer matters was specified with heads of the Treasury Department, the Arms Control and Disarmament Agency, and the Agency for International Development.

At the Defense Department responsibility for military aid and foreign military sales was consolidated under the supervision of the incumbent in a new post: deputy assistant secretary for military assistance and sales. The president also made plans to provide better control, and a more effective case for proposed funding before the Congress, by separating security assistance from development and humanitarian assistance. This included proposed consolidation under a single authorizing act of the various components of security assistance that had been scattered among various pieces of legislation. These included military assistance, military credit sales, grants of excess military stocks, supporting economic assistance, and the public safety program.[7]

In September 1971 the Pentagon established the Defense Security Assistance Agency. Within six months its director was stressing the need to "harmonize our essential strategic objectives, our general defense posture, and our foreign policy requirements with the resources available to meet our security and domestic needs."[8] One can discern in this language concern for the impact of arms transfers on the stocks available for United States forces, a concern that was to intensify over the years of the Nixon administration within the Department of Defense and lead that entity to play a far more restraining role than is generally supposed when it came to responding to foreign requests to purchase arms. As a further means of gauging the impact of arms transfers on readiness of American military elements, planning for the FY1973 military assistance and credit sales programs was done within the Defense Department's planning, programming, and budgeting system, the first time it had been so integrated.

In determinations on arms transfers there was also the usual impact of internal bureaucratic contention, both within and between departments. At State the Bureau of Politico-Military Affairs (PM) continued to play a very major role, one viewed by many as the dominant one, with an effect far greater than that of the more visible office of the undersecretary for security assistance. Especially was this the case when key ad hoc arrangements such as the Vietnam Task Force and the Middle East Task Group were in full swing, with their premium on operational experience and quick response, qualities in which PM possessed a distinct competitive advantage. Policy Planning and the regional bureaus were also often very influential on arms transfer matters where they perceived their interests to be involved.

As a result, pointing out that nearly all conventional arms transfers are the result of government-to-government transactions or other means of control exercised by governments of the states involved, the Arms Control and Disarmament Agency characterized U.S. arms transfers as taking place within "what is probably the most complex legal and regulatory framework of any major arms supplier."[9] Coupled with provisions on "end-use control," which required the recipient of American arms to obtain U.S. consent before transfer of the arms to any other party and which have probably been more rigorously enforced by the United States than by other arms exporters, these procedures and provisions served to focus arms transfer decisions as deliberate instruments of policy. In the view of a number of those who had worked on arms transfer matters from a variety of perspectives, the very complexity of the system for policy determination acted as a "cautionary governor" on U.S. arms transfers, and the further involvement of the Congress typically served to reinforce the influence of the normally "very cautious" permanent bureaucrats in the Departments of State and Defense.[10]

The process of exercising increasingly tight control over all aspects of military equipment transfers continued throughout the Nixon/Ford administrations. Of particular importance was development of a more extensive data base that would permit better assessments of the impact on U.S. force posture and readiness of projected arms transfers. A consolidated document was prepared that, along with planning and programming worksheets relating to various countries, reflected the funded production and maximum possible production of various items of military equipment and the production excess (that which could be made available for sale). These data were used to project the impact of various contemplated sales, variations in timing, and the effect on U.S. forces. The comments of many participants reveal that a major motivation for development of this more detailed management tool was increasing concern within the Department of Defense about the adverse effect on U.S. readiness of arms transfer agreements being approved at higher levels.

Using these mechanisms, the administration set about using arms transfers as a primary instrument of policy.

4. Middle East Policy Objectives

The essence of the story of the Nixon arms transfer policy is in the Middle East.[1] It is where the bulk of the arms were sent. It is where the crucial decisions as to policy were directed and where they represented the most dramatic changes from past policy. And it is where the most spectacular successes were achieved.

What were the goals and achievements of the Nixon arms transfer policy in the Middle East? They were to peel Egyptians away from the Soviets, engineer essential military balance in the Arab-Israeli confrontation, negotiate steps toward agreement on a peace settlement, and provide the prospect for the Egyptians of an alternative source of necessary military hardware and supporting services. Thus it had three genuinely significant goals: to diminish prospects of further major armed conflict between Arabs and Israelis, to remove the Soviets from the region and deny them access to military base facilities there (thus both reducing chances of superpower confrontation in the area and restricting Soviet power projection capabilities), and to make more secure continued U.S. access to the oil it and its allies required by removing the cause for another embargo and lessening Soviet means of interfering.

Further objectives were to establish Iran and Saudi Arabia as major military powers in their own right, able to provide stability in the region and security for the oil there and for the routes of movement of that oil to Western markets. This would sustain conservative governments, pro-Western and now primarily dependent upon the United States for arms and, more especially, for the technical and training assistance they would require to be able to maintain and operate those arms, which in turn would form the basis for their regional power. Those states would then be able to protect the oil resources, their own and those of the region more generally, against other regional threats and to put up a sufficiently good fight with the Soviets or their surrogates to prevent a fait accompli and provide time for other major powers to intervene. The result would be powerful influence in the region opposed to any radical government coming to power in a key oil-producing state, which could endanger Western access to the oil supply. And finally the outcome would successfully compensate for withdrawal of the British stabilizing influence, preempt Soviet moves to take advantage of that change, and avoid having to take over the role at first hand. These are the dimensions of the accomplishments.

Not all of this came about strictly by design, of course, nor would the Nixon administration claim that it did. The essence of its strategic approach was to understand the changing patterns of international politics that it saw taking place, to define American goals in relation to the change, and to devise and carry out policies that would achieve those goals.[2] This meant, as President Nixon liked to illustrate by quoting from the poetry of Chairman Mao, having the perceptiveness and the courage to "seize the day, seize the hour."[3] In the Middle East, only frustration resulted from the first several years of attempts to arrange movement toward a settlement. Some necessary preconditions came to pass, however: Nasser was succeeded by Anwar el-Sadat as president of Egypt. Some channels of communication with him were established by the United States. Another war broke out, during which those channels were kept open and following which a fairly stable military balance pertained between Arabs and Israelis. And the Soviets were fatally slow to make up Arab losses in that war while the United States poured more than $2 billion in replacement equipment into Israel. Finally the oil embargo provided a necessary corrective to the one-sided view of the Middle East political situation that had previously been held by many Americans.

Thus the stage was set for the administration to seize the hour, and seize it they did. Even during the course of the 1973 war, communications with Egypt made it clear that the United States did not want the result to be dominance of the Israeli side (or of course of the Arabs). In the aftermath of the war this channel proved invaluable in negotiating an end to the oil embargo. From there it was a slow, difficult, frustrating, irregular path to further negotiations between the parties concerned at first hand, but progress was made, and the prospects of peace became undeniably brighter. Meanwhile the Soviets were out, the Iranian and Saudi military establishments became stronger, building supporting infrastructures to handle the large quantities of military equipment ordered for future delivery, and the oil continued to flow.

Some have argued that the higher price of oil ought not to have been tolerated, that it should somehow have been averted, and in any event that the arms buildup in the region—which they opposed—could not have taken place unless the escalated oil revenues had made it possible. Only the latter point is valid, and its meaning is not what they suppose. For this, too, represented a necessary precondition for the success of policy in the Middle East. Only when the oil-producing nations were able to obtain a high return on their natural resources, a development not unrelated to the 1973 war and the other events described, were they able to afford to purchase the military wherewithal that gave them the independent capability to safeguard the oil and regional routes of communication, especially given the domestic climate then prevailing in the United States and its impact on foreign aid and security assistance. It is clear from analysis of the congressional and domestic political attitudes that had evolved over the decade beginning with the mid-1960s that there would have been no support for providing on any but a cash-and-carry basis the quantity and quality of arms that would permit Iran and Saudi Arabia to play the regional role envisioned for them.

Thus the enforced provision of the funds to buy weapons, funding resulting from coordinated action on the part of the suppliers of oil to obtain a return that was vastly higher than in the past, was the final essential precondition to success in achieving United States goals in the Middle East. Since the success of the policy depended upon regional powers playing strong and independent regional roles, roles that depended in turn upon their possessing viable military forces, and since it was clear that the United States was not in a position to give them those arms (nor did it want them to have to go to the Soviets to obtain them), the only workable alternative was for regional powers to derive from sales of their oil the necessary revenues to buy what they needed for themselves. This was in fact what happened.

It is extremely significant that, within two weeks of the close of the 1973 Middle East war, Egypt and the United States reestablished diplomatic ties that had been severed since 1967. The achievement this represented for United States diplomacy can be appreciated only when it is remembered that within the preceding month the United States had provided a massive air- and sea-lift of military equipment to Israel, sufficient to provoke the oil embargo, and thereby prevented the Arabs from achieving a decisive military victory. But the achievement was far from accidental and in fact was built on a carefully crafted series of initiatives that reached back even before Richard Nixon was inaugurated.

William Scranton had been dispatched by the president-elect on a mission to the Middle East, where he met with President Nasser of Egypt. Nasser in his turn had signaled a certain receptiveness by sending congratulations to Mr. Nixon on his election.[4] Once in office, the president turned immediately to problems of the Middle East, requesting a study which was the second in the new series directed by National Security Study Memoranda under the National Security Council system supervised by Henry Kissinger. That study was discussed at the first meeting of the National Security Council under the new administration, less than two weeks after the inauguration. The president chose the option calling for intense U.S. involvement in attempting to arrange a peaceful settlement in the region, and more studies quickly followed as to various details of policy options.[5] Within a month Soviet Ambassador Dobrynin paid his initial call on the new president, who told him that "history makes it clear that wars result not so much from arms, or even from arms races, as they do from underlying political differences and political problems."[6]

The approach to the Middle East from the first was concentrated on trying to deal with those underlying problems, which included of course the continuing Arab-Israeli dispute, but also the recent British decision to withdraw their military presence east of the Suez Canal, the growing expansion and deployment of Soviet fleet elements, the rapidly increasing importance of imported oil for the United States and its principal allies, and the imperatives of the domestic and external situations that had led to formulation of the Nixon Doctrine.[7] The president later recalled, and the events bear him out, that what he was trying to do was "to construct a completely new set of power relationships in the Middle East—not only between Israel and the Arabs, but also among the United States, Western Europe, and the Soviet Union."[8] This meant, first of all, halting Soviet domination of the Arab region, and for this it would be necessary to broaden American relationships with the Arab countries. King Hussein of Jordan visited the United States in April 1969, and the president told him that he was "deeply troubled because the absence of diplomatic relations with some of the governments in the Middle East precluded our playing a constructive role in the region."[9] When he met the following day with Mahmoud Fawzi, a representative of President

Nasser, the president told him also of his regret that the United States lacked formal diplomatic ties with Egypt.[10]

Secretary of State Rogers was in primary charge of Middle East affairs during the early days of the new administration. In an address in late 1969 he explained that United States policy was "to encourage the Arabs to accept a permanent peace based on a binding agreement and to urge the Israelis to withdraw from occupied territory when their territorial integrity is assured as envisaged by the Security Council resolution."[11] This basic approach in the Arab-Israeli confrontation of "territory for security" was to hold throughout the administration's efforts to deal with the situation. Shortly thereafter the president forwarded to the Congress the first of his annual reports on foreign policy. Looking toward the Middle East, he observed with candor and accuracy that "the outside powers' interests are greater than their control."[12] Arms transfers were to become a crucial factor for the United States in gaining greater control, then using that control to defuse the prospects of a superpower confrontation and promote the possibilities of a negotiated peace. First, of course, many preliminary events had to transpire, and before things were to get better they first became worse, but the policy was adequate to taking advantage of the opportunities thus presented, so that what at the outset looked to be clearly impossible to many came to be first a possibility and then a reality.

The intention, however, was not from the first to become the arsenal for both sides, nor could it be, for or course the Egyptians were still a Soviet client. At this point, then, the president observed that we had "urged an agreement to limit the shipment of arms to the Middle East as a step which could help stabilize the situation in the absence of a settlement. In the meantime, however, I now reaffirm our stated intention to maintain careful watch on the balance of military forces and to provide arms in friendly states as the need arises."[13] Subsequently the president expressed the view that the "grim distinction" of being our most dangerous problem must go to the Middle East, and again he stated his

concern that the Arab-Israeli conflict had caused disruption
of normal American relations with the Arab states. This was
particularly unfortunate, he felt, because it tended to increase Arab dependence on Soviet support, creating a "dangerous vulnerability to excessive Soviet influence."[14] Thus,
as early as the second foreign policy report, in which these
views were set forth, there were broad public indications of
the necessity to diminish Soviet influence derived from military support, with the clear implication of the possibility of
there being some alternative means of obtaining such support, some other source of arms.

The president understood, as he publicly stated, that
"maintaining the military balance . . . is not by itself a policy
which can bring peace."[15] What was required was "an immensely intricate diplomatic plan designed to gain influence
with the Arab states in order to make the United States a
credible mediator and to reduce Soviet leverage in the region."[16] The arms transfer policy for the Middle East employed by the administration must be evaluated in the
context of the major policy revision for the area, which took
place under the Nixon guidance. From an inherited policy
that strongly favored Israel, Washington came around to a
far more genuinely "evenhanded" policy, one that still supported and sought to help achieve Israeli aspirations but was
also sympathetic to the position held by Arab moderates and
sought to build some bridges of understanding so as to enhance the prospects for achieving a lasting peace agreement.
The view was that in the longer term Israel's interests could
be served only in this way, for periodic armed confrontations
were bound eventually to use up its substance and lead to its
downfall. Not only was it in Israeli interests to create the
conditions that could make possible a negotiated settlement,
but the entire complex of problems in the region could not
be dealt with in any other way. It was not just Arabs against
Israelis, after all—important as that was in global as well as
regional terms—but Arabs against Soviet penetration and
domination, Arab moderates against Arab radicals, oil-
producing Arabs economically against oil consumers,
Soviet proxies against U.S. proxies, United Nations forces

against violence, and terrorists against practically everybody.

In February 1973 the president had entered in his journal the observation that "we just can't let the thing ride and have a hundred million Arabs hating us and providing a fishing ground not only for radicals but, of course, for the Soviets."[17] But first, of course, there was to be another war. In the midst of it, keeping a long-standing commitment to his military deputy, General Alexander Haig, Henry Kissinger —who had just been announced as winner of the Nobel Peace Prize—appeared at the annual banquet of the Association of the United States Army. It was, in fact, the George Catlett Marshall memorial dinner, and it was just twenty years since General Marshall had himself received the Nobel Prize for peace. It was also the twelfth day of the October Middle East war. Before coming to the dinner Dr. Kissinger had met at the White House with four Arab foreign ministers. In the Middle East the Arab states picked that day to proclaim their oil embargo. Thus it seemed especially significant when Kissinger, who had been in office as secretary of state for less than a month, softly observed that "while every moment in history is unique, a few moments are similar in their openness to fundamental choice." The choices and the dangers were "clouded in ambiguity," he said, and to deal with them we would have to summon an "unaccustomed ability to deal in nuance." Within three weeks he was off to the Middle East for the beginning of what was to become known as "shuttle diplomacy." The careful utilization of arms transfers—to Israel, Jordan, Saudi Arabia, Iran, even in prospect to Egypt—was an essential part of the diplomatic initiatives that followed, as indeed it had been from the outset of the administration's efforts to resolve the convoluted problems of the region. In terms of both the supply of arms and their judicious withholding, the nature, quantities, timing, and context of the arms transfers were key variables, requiring balance and delicacy. And it worked.

There is no need to overstate the case for these accomplishments. Certainly they could only have come about due to fortuitous circumstances which, as we have observed, the

Nixon administration did not cause but which it exploited. And there were other events, perhaps unanticipated and surely not wished for, that were turned to advantage by the way in which the administration responded to them and their aftermath, most notably the resumption of war between the Arabs and Israel in 1973. Yet it is no criticism of diplomacy to observe that it did not control all things or that it took advantage of developments that it did not plan or expect—that, after all, is what diplomacy is all about. Were control absolute, diplomacy would be unnecessary, and simple dictates could replace the complex of negotiation and persuasion.

"Ripeness is all," once wrote Shakespeare. This was an insight that clearly informed the Nixon policy toward the Middle East and, in particular, the extended and delicate series of events that led in due course to the shift of Egypt from a role of dependency upon the Soviet Union to that of a more independent actor with important ties, not the least of which were arms transfer ties, with the West. And the use of arms transfers—the withholding, delivery, promise, or hint of the military wherewithal and supporting equipment and services viewed as so important by the states of the region—was the principal instrument of policy that enabled these successes to be attained.

5. Egypt as the Key

No arms were transferred to Egypt by the Nixon administration; yet it is not an exaggeration to say that Egypt was the key to the success of the Nixon policy in the Middle East and to the important part that arms transfer manipulation played in achieving that success.

At the beginning of the 1950s the major arms suppliers to the various nations of the region were the United States, the United Kingdom, and France. In May 1950 these three announced a Tripartite Declaration, which had as its purpose avoidance of an arms race in the area. They also created a Near Eastern Arms Coordinating Committee designed to maintain a rough military balance in the region. For half a decade this approach was in general successful, but the arrangement was upset in late 1955 when Egypt began receiving arms from the Soviet Union by way of Czechoslovakia. Once there was an alternative source of supply, the control employed by monopoly suppliers to restrict the flow of arms was no longer possible.[1] The Soviet-Czech arms deal with Egypt was the first ever concluded with a nonaligned state by the communists,[2] and in addition to introducing the Soviet Union into the middle of the Arab-Israeli conflict it began a long period of primary reliance upon the Soviets for arms on the part of Egypt. One commentator was later to observe that "nowhere have the limits of Soviet policy been more sharply revealed" than in the Soviet-Egyptian relationship,[3] but this was far from evident at the outset of the relationship, and indeed far from self-effectuating thereafter.

Initially Soviet arms for Egypt were apparently intended to dissuade that country from joining the Baghdad Pact and to provide a foothold in the region for Soviet presence. It was not until after the sound trouncing received by Egyptian military forces in the 1967 war that the Soviets, arming the Egyptians once again, were granted the naval base privileges they desired.[4] Significantly, this round of reequipping also meant upgrading of Egyptian military inventories, since some of the systems they had previously held were no longer being produced by the Soviets, so that T-34 tanks and MiG-17 fighters, for example, were replaced with not only newer but more capable systems.[5]

At the end of June 1967, immediately after the conclusion of that war, Soviet President Nikolai Podgorny visited Egypt. He brought with him Marshal Matvei Zakharov, chief of staff of the Soviet army. In addition to pressing for the fleet facilities they had been seeking, the Russians expressed anger that some of the sophisticated equipment they had previously furnished the Egyptians had been captured in the war by the Israelis and had subsequently found its way into the hands of the Americans. To Egyptian officers asking for fresh arms, Zakharov is reported to have shouted: "Arms? What do you need arms for? To deliver these to the Israelis too? What you need is training, training. Then we will see about arms."[6] And training they were to get, with Zakharov himself staying in Egypt indefinitely to take personal charge of Russian assistance in rebuilding the shattered Egyptian armed forces. In an event of equal significance, the oil-producing countries met shortly thereafter at the Khartoum Conference, where they agreed to give financial support to the countries that were bearing the brunt of Israeli military action—initially Egypt and Jordan. King Faisal began by chipping in $50 million.[7]

Other significant events that would reverberate through the next decade also occurred in 1967. President Nasser met with Yasir Arafat, leader of the Palestine Liberation Organization, asking him: "But why not be our Stern? Why not be our Begin? You must be our irresponsible arm. On this basis we will give you all the help we can."[8] When that help got to be more than the Egyptians could handle by themselves, Nasser took Arafat along when he visited the Soviet Union in July 1968. It was the first trip to Moscow for the PLO leader; Anwar Sadat was also in the party. Arafat asked for arms, and subsequently Nasser was brought a letter by the Soviet ambassador indicating that the Soviet Central Committee had decided to allocate $500,000 worth of arms to the Palestine guerrillas.[9]

The overtures to the Soviets caused continuing problems for the Egyptians, not least because of Saudi Arabian distaste for the Russians and the dependence of the Egyptians on financial assistance from the Saudis, for purchase of arms among other things. The Saudis, furthermore, were impatient for more action against Israel. When an Arab summit conference was proposed in 1969, King Faisal was opposed but advised the Egyptians that he would attend on the condition that Egypt "declares openly that it has abandoned all efforts to achieve a peaceful solution, is withdrawing its cooperation with the mission of Dr Jarring and its acceptance of Resolution 242, and is prepared immediately to declare a Jihad [holy war for Islam]."[10]

What came to be called the War of Attrition, which extended through 1969 and 1970, intensified the problems for Egypt. On the one hand it was being pressed to take action against the Israelis, and on the other it was continually disappointed in its expectation of arms from the Soviets, who seemed always to promise more than they delivered. The Egyptians sought to confine the hostilities in their territory to the region of the Suez Canal, but Israeli deep-penetration air strikes, which they were unable to interdict, were both politically and physically punishing. Faced with a situation that was militarily and politically untenable, President Nasser flew to Moscow on 22 January 1970 to seek military assistance.[11]

According to the account rendered by Mohamed Heikal, there was a dramatic confrontation in which Nasser demanded antiaircraft missile batteries and interceptor aircraft, then pointed out that the Egyptians would require extensive training in their employment for which there was no time, the result being that the Soviets would have to provide the initial manning. When Brezhnev observed that this would be a step with "serious international implications," Nasser reportedly responded that if Russia were not prepared to be as forthcoming in its help to Egypt as the Americans were with Israel, he had no choice but to return home, resign his presidency, and turn over direction of his country to a leader who could enlist the help of the Americans. The Soviets quickly adjourned the 10:00 a.m. session and called a special meeting of the Politburo. By 6:00 p.m. the conference with the Egyptians had been reconvened, and the Soviets agreed to meet Nasser's demands. The first Russian MiG-21 aircraft were on site by early April.[12] The Soviet direct operational role eventually led to there being some 12,000 Russian military personnel in Egypt, with thousands of them engaged in operating SA-2 and SA-3 antiaircraft missile batteries along the Suez Canal and 3,000 Soviet instructors working with Egyptian military personnel.[13]

Meanwhile, the Soviets were also trying to get the United States to restrain the Israeli attacks. In his first personal letter to President Nixon, Premier Kosygin on 31 January 1970 complained that all of Egypt was vulnerable to the Israeli air strikes and indicated that the Soviets would be forced to increase arms shipments to Egypt if the United States were unable to confine Israeli combat actions to the immediate area of the tactical battlefield.[14]

That arms buildup had already been promised, of course, and during the first half of 1970 construction continued on the "missile wall" along the Suez, which was eventually to include some 200 batteries. Meanwhile, in April, Assistant Secretary of State Joseph Sisco visited Cairo, meeting with President Nasser. While he did not deny the American commitment to Israel, he told the Egyptian president that the United States had "more flexibility" toward the problems in the Middle East than anywhere else and that what it wanted was a "balanced policy."[15] During the missile construction numerous Egyptian casualties had been taken among the civilian construction workers, with as many as 4,000 killed by one Egyptian account, and President Nasser became increasingly agitated. On 29 June 1970 he flew secretly to Moscow for his second visit of the year, determined to "thrash out the whole question of Egypt's air defences with the Soviet leaders."[16] At the same moment, with an arranged cease-fire about to go into effect, the Soviets and Egyptians surreptitiously moved twelve SA-2 and three SA-3 batteries and other equipment into the Canal Zone in violation of the agreement.[17] Heikal had later written very candidly about this: "Nasser asked me to gain a bit of time for him: he needed six hours so that he could get some dummy missile batteries into position. The Americans would of course photograph from their satellites the exact position of everything at the moment of ceasefire, and Nasser wanted to be able later to replace these dummy missiles with real ones. . . . The Americans were extremely annoyed by the moving of the missiles. They accused the Egyptians of cheating and, to punish them, ostentatiously agreed to supply Israel with more arms."[18]

Perhaps this was one of the reasons why Heikal later observed that "Sadat . . . did have a chance of moving the Americans towards an understanding of Egypt's position that, to be realistic, Nasser had never had. The mistrust between Nasser and the Americans ran too deep."[19] Yet Nasser, who had only a short time left to live, made an important contribution to future developments in late July 1970, when he announced Egypt's acceptance of the United States proposal for a cease-fire and initiation of informal talks among Israel, Egypt, and Jordan. In so doing Nasser was the first Arab leader to depart from the tenets of the Khartoum pact by agreeing in principle to hold discussions with the Israelis. Among the immediate results were suspension of Egypt's subsidies from the Arab oil-producing countries, rioting in the more radical Arab states, and the onset of civil war in Jordan.[20]

On 1 July 1970, in what someone later called "a famous indiscretion," Dr. Kissinger told reporters at San Clemente that "the goal of American policy was to 'expel' the Russians from the Middle East."[21] And that is exactly what was done. In November of the same year Kamel Adham, the brother-in-law of Saudi Arabian King Faisal, was in Cairo on a visit. In conversation with President Sadat:

He talked about the Russian presence in Egypt, saying how much it alarmed the Americans, and pointing out that this was important at a time when the Saudis were trying to get the Americans more actively interested in the Middle East's problems. President Sadat's answer was that Egypt depended on the Soviet Union for so much, whereas the Americans were providing Israel with everything it asked for, to the extent that during the war of attrition they had been able to bomb Egypt for seventeen hours consecutively. The President told Kamel Adham: "I would not only bring in the Russians —I would bring in the devil himself if he could defend me." But he added that if the first phase of Israeli withdrawal were completed he could promise that he would get the Russians out. Kamel Adham asked President Sadat if he could pass this on to the Americans and the President said he could.[22]

Other visitors continued to implement a sort of dialogue-once-removed between the Egyptians and the Americans. In June 1971 King Faisal was in Cairo on his return from the United States. The Egyptians had just signed a fifteen-year Treaty of Friendship and Cooperation with the Soviet Union, and King Faisal told them that he had been closely questioned by the Americans as to his interpretation of its meaning. He had, he told them, said he was sure Egypt would never adopt communism.[23]

It appears that the Soviets were anxious for the treaty, concerned perhaps by the recent dismissal from the Egyptian government of officials who had been closely associated with the Russians. President Podgorny himself brought the treaty to Cairo. It had two clauses that were important to President Sadat, one providing for cooperation in dealing with any threat to the peace and one committing the Soviet Union to supply Egypt with the military equipment it needed, so he signed, on 27 May 1971.[24] From the Soviet standpoint, the treaty required Egypt to formally abandon its stance of nonalignment and involved as well a commitment toward socialism in domestic policy. Egypt's hopes that the treaty would also serve to limit Soviet interference in its internal affairs were to prove unfounded. A later assessment concluded that the Soviet efforts to move from involvement to influence in the region provoked reactions and that "the reaction was most violent in Egypt, where Soviet imperialism had taken its most blatant form, and Soviet restrictions on arms transfers had been especially resented."[25]

President Nixon was asked about the Soviet-Egyptian treaty at a news conference at the beginning of June. He interpreted it entirely in terms of its impact on arms transfers. The treaty, he said, "will have effect only in terms of how it might affect the arms balance. In the event that this will be followed by an introduction of more weapons into the Middle Eastern area, it can only mean a new arms race and could greatly jeopardize the chances for peace. . . . Whether the Soviets follow up with large-scale arms shipments into the area will determine whether or not it increases the chances for peace or sharply increases the chances for war."[26] It is interesting to note that during all of 1971 the president was asked only one question relating to arms transfers at any of his news conferences, and that related to this same subject, concern about Soviet introduction of arms into the Middle East, and came during an atypical televised news conference with only a few selected correspondents present.[27]

President Nasser's last official act had been to preside over an Arab summit meeting held in Cairo in late September 1970 to resolve the crisis in Jordan, which resulted in agreement on a cease-fire and mutual withdrawal of army

and guerrilla forces from the cities where combat had been
in progress. The following day he died of a heart attack. The
frustration and rage he must have felt in the face of repeated
Soviet failures to meet his expectations in providing arms
adequate to the Egyptians' assessment of their needs were
reflected years later in an interview with Anwar Sadat, who
had of course succeeded Nasser as president. Asked about
Soviet favoritism in providing arms to some supposed allies
and not to others, he responded with questions of his own:
"Why did they let Abdul Nasser reach the point of despair
in 1970 and return from Moscow utterly disheartened because of the deterrent weapon [interceptor aircraft] which
we had been demanding from them for three years before
Abdul Nasser's death? We told them that we wanted the
deterrent weapon to face Israel when it strikes in the inte-
rior of Egypt. Why did they let Abdul Nasser die before
giving him the deterrent weapon?"[28] The new president had
his own problems. Nasser had left him "a demoralized Egyptian government with an empty treasury, no prospects of
significant aid and a troublesome alliance with the Soviet
Union."[29]

Nasser's funeral a few days later set some further events
in train. The United States, not having formal diplomatic
relations with the Egyptians, sent as its representative Secretary of Health, Education, and Welfare Elliot Richardson,
who arranged to meet privately during his visit with President Sadat, thereby establishing the first contacts with the
new president.[30] Premier Kosygin was also in Cairo for the
funeral and took the occasion of his visit to seek a meeting
with some of those who would be taking over the Egyptian
leadership. General Mohammed Fawzi, commander-in-chief
of the Egyptian armed forces, brought up the question of
arms, pointing out that the Americans had undertaken a
new arms program for Israel, which included the Shrike
missile and additional Phantom and Skyhawk jet aircraft. It
was, he emphasized, extremely important that "after Nasser's death the Egyptian forces should have full confidence
in the continued flow of Soviet weapons." He could as well

have pointed out that that would be a new experience for them, but apparently refrained from doing so. Zakharov was in the Soviet party and said he would do what he could, but complained that the Egyptian "shopping list" was too big.[31]

In his report on foreign policy for 1969–1970 Secretary of State Rogers emphasized that since 1969 the United States had, both publicly and privately, made it clear to Egypt (the United Arab Republic, at that point) that it was prepared to restore diplomatic relations without any conditions.[32] But he also documented the cease-fire violations of August 1970, asserting that "the Egyptians, aided by the Soviet Union, were beginning a major violation of the military stand-still agreement, during which they constructed numerous new sites for SA-2 and SA-3 missiles, completed many that had barely been started on August 7, moved missile deployments closer to the Canal, and activated more missile units within the zone [fifty kilometers east and west of the cease-fire line]."[33]

Things were also not going so smoothly with the Soviets for the new Egyptian president. Although Soviet President Podgorny had visited Egypt in January 1971 and had told the chief of staff of the Egyptian army that the arms he had requested "would be forthcoming,"[34] by March President Sadat was sufficiently concerned to make his own secret visit to Moscow. He wanted three things: a joint strategy, equal footing with Israel as far as arms were concerned, and improvement in the current flow of arms deliveries.[35] Despite the efforts by Sadat, Heikal reported that "trouble was brewing with the Soviets. They were ... suspicious of the new regime, while on the Egyptian side mistrust had been generated by the demand for a naval base and by difficulties over arms supplies."[36]

In May Secretary of State Rogers was in Egypt to confer with President Sadat, stressing the idea of "withdrawal for guarantees." It was pointed out to him that Egypt had accepted the latest proposals for written commitments to peace and withdrawal, based on Resolution 242, that the Israelis had declined to do so, and yet the United States still continued to supply them with arms. According to Heikal's

account, Secretary Rogers responded that the United States
wanted peace but was unable to put pressure on Israel. Egyptian Foreign Minister Riad suggested that "the only effective
form of pressure on Israel would be an American embargo
on arms."[37]

Looking at it from the other side, American concern over
the Soviet supply of arms to Egypt, and the prospect that the
situation might be building toward a renewal of hostilities,
was reflected in the president's foreign policy report of February 1972. The superpower implications were dealt with
first, as the president said that "the Soviet Union's effort to
use the Arab-Israeli conflict to perpetuate and expand its
own military position in Egypt has been a matter of concern
to the United States. The USSR has taken advantage of
Egypt's increasing dependence on Soviet military supply to
gain the use of naval and air facilities in Egypt. This has
serious implications for the stability of the balance of power
locally, regionally in the Eastern Mediterranean, and glob-
ally."[38] Recapitulating the Soviet military material provided to the Egyptians during 1970, and not forgetting their
complicity in violations of the cease-fire agreement, Mr.
Nixon added that "the Soviets since that time have introduced into Egypt SA-6 mobile surface-to-air missiles and the
FOXBAT and other advanced MIG aircraft. Most recently
they have reintroduced TU-16 bombers equipped with long-
range air-to-surface missiles. Much of this equipment was
operated and defended exclusively by Soviets."[39] It was clear
that United States policy contemplated somehow reversing
the trends reflected in the "increasing dependence" of the
Egyptians on the Soviets for arms. As the president observed, "the essential problem of peace" is "to shape new
patterns of order." To this end, "new relationships must be
shaped."[40]

One such relationship, although it would have to wait
until after the catalytic effects of yet another Middle East
war, was clearly that between the United States and Egypt,
and just as clearly it was one that would imply an eventual
military supplier relationship as an essential component,

indeed the key element enabling Egypt to cut itself loose from such dependence upon the Soviet Union. In this the USSR was to be an unwitting accomplice, for no matter the American concern about the buildup of arms being provided the Egyptians by the Soviets, in Egypt there was dissatisfaction with the quantity, quality, and timing of the supply of arms, and this was a concern and indeed the primary factor in the growing disenchantment with the relationship on the part of President Sadat and his government. Thus, "from 1972 onward, Soviet witholding of certain aircraft and missile types desired by Egypt, and the slowdown in delivery of other agreed shipments altered relations. These imposed constraints were a major factor in President Sadat's dismissal of Soviet advisers in July 1972, in his later dramatic shift to resume political relations with and reliance on the United States, and in his decision to diversify arms supply by purchase in Europe or, if possible, from the United States."[41]

Mohamed Heikal confirms that "above all there were the arguments about arms supply which were the main theme of all four visits the President paid to Moscow between the time he took over and the end of April 1972."[42] As another analyst later suggested, "such constraints on arms supply had been decisive in shaping Egyptian behaviour,"[43] but surely not in the way the Soviets must have contemplated. Events rapidly built to a climax. On 1 June 1972 President Sadat sent an urgent message, with a follow-up message two weeks later. Brezhnev's reply of 7 July completely ignored the request.[44] The night before receiving that reply, Mr. Sadat had told his vice-president, Dr. Mahmoud Fawzi, that "he was thinking of asking the Soviet Union to withdraw its military personnel from Egypt because they had become a burden."[45] The day after the message from Brezhnev, Sadat sent for Soviet Ambassador Vinogradov and informed him that he had reached such a decision, to take effect 17 July 1972.

President Sadat followed up the expulsion of the Soviets with diplomatic initiatives aimed at the United States. An

official of the Department of State summarized their essence: "You said you wanted us to expel the Russians and
now we did it. . . . So what are you going to do to help us?"
"It was clear," said a senior official of the department, "that
Sadat was taking the first step to engage the United States
in active diplomacy."[46] President Nixon sent several mes-
sages to President Sadat through intelligence channels via
his national security advisor, Hafez Ismail, bypassing the
regular diplomatic means. Acknowledging that the expulsion was an important step, Mr. Nixon pledged to give priority attention to the problems of the Middle East once
presidential elections had been completed and the negotiations over Vietnam were wrapped up.[47] Early the next year
Dr. Kissinger met with Mr. Ismail in the United States in
three secret sessions over a two-day period, but whatever
progress there may have been was largely negated by Ismail's reading while en route back to Egypt that the United
States had decided to make available to Israel an additional
shipment of forty-eight Phantom and thirty-six Skyhawk
aircraft. In one analyst's view, "the abortive Ismail mission
was the turning point on the path to war."[48]

But there were other pressures on the Sadat government
that made it extremely difficult to be patient while the
United States wrapped up other, more pressing business.
The expulsion of the Soviets had increased Sadat's standing
with the Egyptian public, added to his prestige with the
Egyptian military establishment, and put him in better
standing with other Arab governments.[49] But those who had
been bankrolling Egypt wanted some action for their money,
and the action they wanted was a military campaign to
regain lost Arab territory being occupied by the Israelis.
The problem was to obtain the needed military wherewithal to mount an offensive with a reasonable chance of success.

The solution to this problem appeared in the near magical
effect of the expulsion upon the customarily dilatory Soviet
supply of arms. Under the influence of what some observers
claimed was pressure from hard-liners in his own govern-

ment, Mr. Brezhnev decided to open the floodgates and let
Egypt have the sophisticated equipment it wanted. After a
flurry of activity in which officials shuttled back and forth
making the necessary arrangements, large quantities of
Russian tanks, aircraft, bridging equipment, and electronic
gear started pouring into Egypt.[50] Major deliveries had
begun by the end of 1972, and Heikal was later to report that
"between December 1972 and June 1973 we received more
arms from them than in the whole of the two preceding
years." At one point the response became so copious that
President Sadat told an advisor, "they are drowning me in
new arms."[51] This must have been gratifying indeed, for the
president had previously told a group of editors that the
main factor in his decision to call a halt in Egyptian relations with the Soviets had been the "evasion and delay over
the delivery of arms to which . . . he had been continuously
subjected."[52]

The year 1972 represented the peak of optimism regarding Soviet-American relationships. The Moscow summit, the
signing of the SALT agreements on ABM systems and offensive weapons and a cluster of other agreements for cooperation in various realms, the articulation of a set of basic
principles to govern the superpower relationship, the influence on this relationship of the United States' opening to the
People's Republic of China, diminished tensions in the Middle East (or so it appeared), the prospects of increased trade
—all led to a perception that détente was developing substance to match the rhetoric and that a new era lay ahead.
All of this was to change, and dramatically, within a few
more months, principally because of the massive Soviet involvement in Egyptian actions leading to yet another war in
the Middle East. It is one of the major ironies of recent
diplomatic history that this led not only to more realistic
appreciation of the inherent inability to cooperate to mutual
advantage that crippled the Soviet approach, but to a chain
of events that resulted in loss of all the previous Soviet advantage in Egypt and much of that in the Middle East more
generally and opened the door for U.S. influence and the

much-improved prospects of a viable settlement of regional confrontations.

The president's fourth annual foreign policy report was published in May 1973. It included a very direct hint to Egypt in an observation that "the United States considers it a principal objective to rebuild its political relations with those Arab states with whom we enjoyed good relations for most of the postwar period but which broke relations with us in 1967. . . . The United States is prepared for normal bilateral relations with all the nations of the Middle East."[53] When war erupted, noted the president in his memoirs, "we had a particularly delicate situation insofar as the Egyptians were concerned. Beginning in February 1973, with a view toward building better relations, we had had a series of private contacts with them. . . . I hoped that we could support [the Israelis] in such a way that we would not force an irreparable break with the Egyptians, the Syrians, and the other Arab nations."[54]

Egypt had "come to accept the idea of a limited attack aimed primarily at opening up political possibilities."[55] Anthony Nutting, who had served as foreign minister under Anthony Eden, once closed a book on the Middle East with these observations: "It is almost as impossible to finish a history of the Arabs as it is to forecast their future. . . . [T]he Arabs, like, probably, no other people in the world except the Irish, are irrational and emotional to a point where they think only with their hearts, never with their heads. . . . No race on earth will more eagerly or cheerfully cut off its nose to spite its face."[56] While some allowance must be made for the adverse impact on one's perceptions of the experience of Suez in 1956, the attitude nevertheless parallels the reaction displayed by many to the Egyptian crossing of the Canal on 6 October 1973, thereby initiating the fourth major war in the Middle East in a quarter century. As it turned out, however, a large number of political possibilities were opened up by this bold adventure—not all of them, of course, equally attractive from the standpoint of the United States. Dr. Kissinger acknowledged this when he met *Al Ahram* editor Mohamed Heikal at dinner at the home of Egyptian Foreign

Minister Fahmi on 7 November 1973, an event that just by its occurrence is enough to stagger the imagination: "The Soviet Union can give you arms," Heikal recalls Kissinger having told him, "but the United States can give you a just solution which will give you back your territories, *especially as you have been able really to change the situation in the Middle East.*"[57]

Two objectives guided U.S. actions from the onset of hostilities: get the fighting stopped as quickly as possible, and get it stopped on terms that would be most conducive to a more permanent settlement in the region. As a practical matter, these objectives were not necessarily fully compatible at every turn. Furthermore, the range of possible outcomes caused painful dilemmas for U.S. policy makers. The prospect of an oil embargo, for example, was not entirely unanticipated. Sheikh Ahmed Zaki al-Yamani, the Saudi minister of petroleum, had been sent by King Faisal on a special mission to Washington the preceding April. His purpose was to point out to the administration that Saudi Arabia was not interested in increasing production to meet U.S. needs unless there was some indication that the United States was working to bring about the withdrawal of Israel from the occupied territories. Yamani met with a number of top officials, including Kissinger, Secretary of the Treasury George Shultz, and Secretary of State Rogers. However seriously his message may have been taken, the Americans apparently felt they were in no position to make any commitments, for Yamani was later quoted as having said it was "a dialogue of the deaf."[58]

Looking back from the vantage point of five elapsed years, a financial analyst assessed the oil embargo and the manyfold increase in oil prices that followed. They constituted, he wrote, "one of the economic watersheds of the modern era." In their wake came "a deep, worldwide recession and an upheaval in the balance of world power between oil producers and consumers as OPEC, at a stroke, demonstrated it held the power to strangle the economic well-being of the Western World by the way it priced and supplied its oil."[59] We have suggested that the major increases in oil prices served an essential purpose in providing to the Arab states the wherewithal to become independent in political and arms-supply terms, an aspect of the developments in the Middle East that has received too little attention. But, given the cataclysmic effects as viewed by others, many critics have alleged that the United States failed to recognize—until it was forced to do so by the 1973 embargo—that it and especially its allies were increasingly dependent upon oil from the Middle East, with attendant strategic implications. The record shows otherwise. Thus the president's foreign policy report of May 1973:

> Our own requirements for Persian Gulf oil have been small—about ten percent of our total oil imports—but they will rise as U.S. energy demand expands. Assurance of the continuing flow of Middle East energy resources is increasingly important for the United States, Western Europe, and Japan. . . .
>
> As for the relations between producer and consumer nations, here too we believe there is a shared interest. We both stand to gain from a stable and reliable economic relationship, ensuring revenues for them and energy resources for us. Oil revenues paid to Persian Gulf states have trebled in the last five years, financing their economic development and providing an expanding market for us.[60]

An interesting account of plans for use of oil supplies as a factor in the 1973 war has been provided by Mohamed Heikal. In the spring of 1973 Mustafa Khalil, former Egyptian deputy prime minister for industry and mineral resources, was invited to undertake "a study of the energy crisis in the United States and its implications for the Arab states" at *Al Ahram's* Centre for Strategic Studies. The Khalil Report was finished on 1 October 1973. Five days later its author was asked to prepare a memorandum for President Sadat on how oil should be used in the war. On 8 October

Khalil and Sayed Marei, assistant to the president, were sent to Saudi Arabia and the Gulf states to explain Egypt's attitude to the war. "The point of view expressed in the Khalil report was that oil should be regarded as a strategic and economic commodity and not as a weapon in itself, since oil by itself can never win a war." Thus, "if the Arabs are supplying any country with one strategic commodity—oil—and getting another strategic commodity—armaments—in return, they have a right to expect that the flow in both directions will continue whether there is a war with Israel or not. If the arms are withdrawn, then the oil would have to be withdrawn too."[61]

The Egyptians thought they had King Faisal convinced that this was the correct approach to the use of oil as an adjunct to the prosecution of the war. Subsequently the Arab oil ministers met in Kuwait. But, instead of the carefully coordinated policy the Egyptians had hoped for, there emerged only "a staggering increase in price and a moderate cut in production," which the Egyptians thought served no one's interests better than those of the American oil companies.[62] From then on, according to Mr. Heikal, both President Sadat and King Faisal were bombarded with cables and letters from President Nixon and Dr. Kissinger stressing the disabling effect of the oil embargo on their efforts to deal with the American Congress, public opinion, and the press, much less the Israelis. "Eventually," he observed, "they felt obliged to comply, but before any of their aims had been achieved."[63] From the standpoint of an American observer, and one not a supporter of the president, it appeared that "Richard Nixon's stubborn policies in the Middle East paid off when the Arab oil producers finally lifted their embargo, on March 18."[64]

Besides the obvious and highly unpalatable possibility of being slapped with an oil embargo, numerous other difficulties presented themselves to the administration as it attempted to deal with the sudden onset of war in the Middle East. As always, concern for the strategic implications of a crisis was paramount, with uncertainty as to what the

Soviets were doing and might do an element that persisted throughout the period of active fighting, culminating in the much discussed and debated presidential decision to place U.S. forces worldwide on an increased alert status against the possibility of unilateral Soviet intervention in the theater. The question of providing arms to the Israelis was from the first an agonizing one. Whereas the Soviets no doubt had their own problems dealing with the demands of their Arab clients, the United States had not only to deal with the early, very substantial, and persistent Israeli demands for arms resupply but with the very active domestic Israeli lobby, elements in the Congress, the press, and the public. All of this was made more excruciating by the quite obvious connection between the resupply of arms and the threat of an oil embargo.

Meanwhile, the Egyptians were enjoying early military success, helped by the element of surprise and by the highly effective umbrella of air defense missiles and late-model antitank missiles the Soviets had provided. From the Egyptian standpoint, the timing had been right. One factor was the prospect that they had received all the arms they could expect, so that they were at the peak of their military capability. Additional incentive was provided by the economic difficulties deriving from the extended period of "no peace, no war." The only source of help was the richer Arab states, and they were not going to give more unless there was some movement against the Israelis. Finally there was concern that the superpowers were, in the context of détente, possibly going to impose a solution upon the region. It might be, therefore, as President Sadat told his National Security Council four days before the attack, that this was their last chance for action.[65]

On the evening of 8 October 1973 President Sadat received a telephone call from Vinogradov, the Soviet ambassador, advising him that a Soviet airlift of arms was shortly to begin. Mohamed Heikal has recorded Sadat's reaction: "Yes, yes. Magnificent! Magnificent! Tell Comrade Brezhnev I feel thankful to him from the bottom of my heart. Tell

Brezhnev that it is Soviet arms which achieved the miracle of the crossing."[66] The ensuing arms buildup on both sides, the course of the fighting, and the manner of its termination can, given the close involvement of the United States with Israel, be more conveniently analyzed from that perspective. What is most significant here—besides the fact that during the course of these hostilities, and even while providing massive resupply of arms to the Israelis, the United States kept in communication with the Egyptians, persuading President Sadat that it would be unwise to have the Soviets move troops into the area unilaterally—was that one month less a day after Sadat's ecstatic telephone conversation with the Soviet ambassador Secretary of State Kissinger and President Sadat met in Cairo, the secretary dined at the residence of the foreign minister, and the United States and Egypt reestablished diplomatic relations. Even before that, on 29 October, Ismail Fahmi had visited the United States, even before Golda Meir arrived in Washington to convey her gratitude for the American resupply of arms. By December, meeting again with President Sadat, Dr. Kissinger could cable the president that the Egyptian leader had promised to get the oil embargo lifted (during the first half of January, a deadline he was not to meet).[67] By March 1974 the oil was flowing again.

Nearly five years later, the *New York Times* editorialized as follows: "In refusing to block President Carter's package of jet fighter sales to Israel, Egypt and Saudi Arabia, the Senate has in effect ratified the contentions of Richard Nixon, Gerald Ford and now Jimmy Carter that the United States can uphold its commitment to Israel and at the same time enter into partnership with its moderate Arab neighbors. In short, the American Presidency has now won political ratification for a long-evolving and complex Middle East policy that aims to promote the security and economic interests of both the United States and Israel."[68]

Shortly thereafter, explaining the significance of the Carter administration's decision to sell sixty F-15 jet aircraft to Saudi Arabia as part of the Middle East arms sales package also involving arms for Egypt and Israel, *U.S. News & World Report*'s deputy editor Joseph Fromm recalled that "actually, the reassessment of U.S. policy in the Mideast began long before these events or the controversy over the sale of jets. The first tentative moves toward what was called 'an evenhanded policy' were made early in the administration of Richard Nixon." The October 1973 war gave added impetus to these initiatives, he reflected, leading Secretary Kissinger to work hard to keep Israel from crushing Arab armies, thereby establishing the United States in the role of "honest broker" in the region. But, as he also recalled, in so doing Dr. Kissinger faced powerful congressional opposition.[69] Two years later, when agreements on partial disengagement had been completed, President Sadat was to tell the United States Congress that his main reason for entering into the agreements, and for accepting the risks they entailed, was that they "provided for unprecedented American involvement in the conflict."[70] The twin themes of skillful diplomacy and effective use of arms transfers run throughout this story. Just as the trust and confidence that had been engendered were to make possible President Sadat's decision to take the risk of entering into agreement, so had the prospect of another source of arms that would enable Egypt to break its dependence upon the Soviets been a crucial factor. But the Soviets had to play their part, too, however unintentionally. For once the conflict was resolved, the Egyptian problems with the Soviets returned, as vexing as ever. When President Nixon made his historic visit to Egypt in June 1974, President Sadat told him with surprising candor of his difficulties in getting the military help they wanted from the Soviets: "We just gave up on them," he concluded.[71]

In December 1974, more than a year after the war had been brought to a halt, Egyptian Foreign Minister Fahmi and War Minister Gamassy visited Moscow, where they met with Secretary Brezhnev to discuss, as President Sadat later revealed, the necessity for replacing arms destroyed in the war and the new systems that were required. Shortly thereafter Brezhnev's planned visit to Cairo, which had been scheduled for January 1975, was canceled, and Sadat announced that the Soviets had refused Egyptian arms requests, complaining that only some spare parts paid for by Algeria had been received from the USSR and none of the "decisive weapons" that Syria had received.[72] Even two years later Sadat said in an interview that "the arms lost in the October War have not been compensated for, nor is the problem of the debts finally settled."[73] Further in the discussion he pointed out that Egypt had thus "followed an open-door policy by diversifying the sources of arms."[74] He took other occasions to publicly criticize Soviet arms transfer policies after the 1973 war. While the Egyptians lacked replacements for their war losses of material, he asserted, the Syrians were oversupplied. "Already, in 1972, President Assad told me he did not know where to put the arms Moscow was delivering," Sadat said. But he viewed Syria as totally dependent upon the Soviets as a result and said that Egypt could have received the same bounty of weapons if it had "accepted to give up its independence."[75]

Keeping their Soviet-built equipment functioning in the absence of Russian cooperation was becoming more and more of a problem for the Egyptians, as they faced a major difficulty confronting every country that shifts its primary dependence from one arms supplier to another, or to a variety of others. The Soviets had apparently taken active steps to intensify the pain, using the provisions of a licensing agreement that barred the sale of equipment to third parties to prevent India from supplying spare parts for MiG-21 fighter aircraft. Egypt was able to get thirty engines for MiGs from the Chinese, who built MiG-17s and MiG-21s on license from the Soviets and were in a position not to care whether they observed the third-party rule or not.[76] Meanwhile, Libya had during 1974 signed an order for $2 billion in arms from the Soviet Union, a step that must have given Egypt pause, considering the ups and downs in the relationship with the Qadhafi regime.[77] Speaking to the Egyptian People's Assembly, President Sadat reflected his reaction to these developments in recounting the "contemptuous treatment" he had been subjected to by Soviet officials. When Sadat abused the Russians publicly like this his advisors cringed, asking,"Where are we going to get arms?"[78]

Other developments were far more encouraging for the Sadat government. In October 1974 an Arab summit conference in Rabat had resulted in a pledge from the oil-producing states to provide funds amounting to well over $2 billion a year for the next four years to the front-line Arab countries —Egypt, Jordan, and Syria—and to the Palestinians so as to finance arms acquisitions, thereby providing a measure of reassurance and a basis for midrange planning that had previously been lacking.[79] Support from the United States was also forthcoming. Testifying in behalf of the administration's proposed foreign assistance program, which included $250 million in economic assistance for Egypt, Secretary Kissinger in July 1974 tied it to Egypt's "bold decision to move from confrontation to negotiation as a means of resolving the Arab-Israel dispute" and remarked that Egyptian leaders had "shown a desire to substitute friendship and trust in the US for the hostility and distrust which has so long divided us."[80]

Only the month before President Nixon had visited Egypt. President Sadat told him, speaking before a large crowd, that he had decided to cast his lot with the Americans. The mutuality of that commitment was demonstrated by the appearances of the two leaders together not only in Cairo, but on a lengthy trip by train to the city of Alexandria. On the third day of the visit the two presidents issued a document entitled "Principles of Relations and Cooperation Between Egypt and the United States." According to a later account, President Nixon had hinted "that eventually Egypt could count on American weapons," while subsequently President Ford and Secretary Kissinger had "suggested to Sadat that the American embargo on arms to Egypt would be terminated soon, however modestly at first."[81]

As early as the beginning of 1974 there was recognition in the Department of State that reestablishment of United States-Egyptian diplomatic relations would alter the arms

embargo policy toward Egypt, Syria, and Iraq that had been
in effect since the 1967 Arab-Israeli war. The 1974 routine
annual review of countries in the Middle East eligible to
purchase defense articles from the United States under the
Foreign Military Sales Act—which required a presidential
determination that such sales would "strengthen the security of the United States and promote world peace"—did not
result in Egypt's being added, there being "no sentiment" in
the department for making any additions at that time. But
it would not be long. The British meanwhile announced their
intention of lifting the arms embargo of the Middle East that
they had been observing.

The Egyptian foreign minister, Ismail Fahmi, had visited
the United States in August 1974. His itinerary included an
appearance on the "Today" show on network television,
where he emphasized that the Egyptians were "building and
restoring relations with the United States." "Did you ask for
any arms?" pressed his interviewer. "No," Fahmi replied.
"We have, as you may know—have started to diversify our
resources of armaments."[82] Secretary Kissinger's approach
to the Arabs at this point was characterized as saying in
effect: "I know what you want—your territory—and I'm
working on it. Meanwhile, I'll give you *everything else* you
want to compete in the twentieth century."[83] One did not
have to look very far to observe that, like it or not, among
those things most necessary for competing was an adequate
supply of modern arms.

On 15 March 1976 Egypt unilaterally abrogated the
Treaty of Friendship and Cooperation with the USSR which
dated from 1971, with President Sadat proclaiming, "I fall
on my knees only before God." The Soviet Union, observed
one analyst, "paid a heavy penalty for the duplicity of her
detente policy in the Middle East."[84] At a year-end interview
President Sadat said with simple finality: "The story of the
treaty, the experts and the facilities is over."[85]

The most pressing strategic concern that had faced the
United States in the Middle East was the prospect of Soviet
hegemony there. Looking back from the vantage of 1978, and

over all that had happened in the region over the preceding half decade, a columnist expressing support for the proposed package deal of arms sales, which included Egypt, could cite as the "worst case" outcome in the region "an Egyptian rapprochement with the Soviet Union," which he called "the most dismaying prospect for Israel and the United States."[86] And that was the essence of it, of course—that it was to the great benefit of both the United States *and* Israel that Egypt be provided a viable alternative to dependence upon the Soviets, and very much to the advantage of Egypt and the Arab world as well. The evidence that arms transfer policy —always in conjunction with patient and balanced diplomacy—was the key to bringing about the transformation is simply overwhelming.

"No one should forget that Hafiz al-Assad and I made the 6 October decision—the most serious decision which affected and will affect the history of our Arab nation for generations to come," reminded President Sadat at the end of 1976.[87] Mohamed Heikal had once reported President Nasser's aspiration to "lift the Middle East dispute from the local to the international level" where it "might be more evenly balanced."[88] If the problems of the region had not been dealt with at the international level prior to the 1973 war, they most assuredly were thereafter. In the course of it, the Congressional Research Service maintained in a subsequent report, "arms sales to the Middle East . . . have perhaps been the major instrument for maintaining American influence in the region."[89]

President Sadat visited Washington in October 1975 and was told that the United States was prepared to offer "certain military assistance" to Egypt.[90] In an interview before the visit, President Ford stated that he believed the United States had "an implied commitment" to make available certain equipment to the Egyptians.[91] By February 1976 the Ford administration had decided to initiate the supply relationship with Egypt by approving a sale involving six C-130 transport aircraft, following up on the previous year's sales of trucks, jeeps, and an aerial reconnaissance camera system. The aircraft sale was characterized in news reports as "an effort to prove to Egypt that the United States is determined to help compensate for Cairo's decision to end its dependence on the Soviet Union as its principal arms supplier."[92] President Ford had foreshadowed this decision, as well as later installments of combat equipment, when he sent his message on military assistance to the Congress on 20 January 1976. "In the Middle East and elsewhere," he wrote, "we must maintain our flexibility to respond to future assistance requirements which cannot now be reckoned with precision."[93] The Carter administration followed through on the "implied commitment," selling the Egyptians fifty F-5 aircraft as part of a package sale including higher-performance aircraft for the Israelis and Saudis, a move that required considerable political courage and determination in the face of congressional opposition.

Yet it is by no means clear that at the time the eventual implications of the Nixon approach to weaning Egypt from Soviet dependence as an essential precondition to peace in the Middle East were widely understood, even within the United States government. Commenting on the Carter administration's proposed package of fighter aircraft sales to the Israelis, Saudis, and Egyptians in early 1978, even so astute an observer as Richard Burt labeled this "fighter diplomacy" and argued that, "prior to Mr. Sadat's initiative [the dramatic peace initiative of November 1977], the Carter administration, like its predecessor, had been able to put off Egyptian requests for American fighters, providing only 'non-lethal' equipment such as C-130 transport planes." The parallel decision to sell F-15 fighters to the Saudis, according to officials cited by Mr. Burt, " 'short-circuited' the work under way by the arms export board." Why that board was not better informed as to the imperatives of continuation of this painfully slow but markedly successful evolutionary approach to peace in the Middle East is hard to understand. White House officials confided, Mr. Burt tells us, that in order to support Sadat President Carter "concluded that he would soon have to plunge ahead with the first American sale of combat aircraft to a so-called 'front-line' Arab state."[94] This "plunge," it would seem from a review of the evolving policy over the course of a decade, was more properly the logical, inevitable, and eminently predictable result of the coming to fruition of the Nixon/Kissinger policy. Far from constituting an undesirable but unavoidable outcome, it was what they had been working for all those years. It was an indispensable means to the end of blocking Soviet hegemony and promoting the conditions under which a negotiated and lasting settlement of the Arab-Israeli confrontation could be concluded.

6. Trying to Live with Israel

If the essence of dealing with Egypt was forging a new relationship, with Israel it was trying to manage an existing one. Even before the Nixon administration took office it was involved in a controversy concerning Israel, and predictably it concerned the supply of arms. The year 1968 had opened with a visit to Washington by the Israeli prime minister, Levi Eshkol, during which a joint communiqué was issued stating that "the United States would keep under active and sympathetic consideration the defensive needs of Israel."[1]

In the wake of that public reassurance there followed a prolonged controversy over whether to comply with Israeli requests to purchase supersonic F-4 Phantom jet aircraft. Following press reports that President Johnson had decided against approving the sale, Vice-President Hubert Humphrey, pressing his presidential candidacy before a convention of the Zionist Organization of America, told his audience to applause that America must do all it could to end the arms race in the Middle East, but until then we must "continue United States military assistance to Israel, including what has become such a symbol and I think also a necessity, the supersonic plane, such as Phantom jets."[2] Richard Nixon, addressing the same audience, had stated that the military balance in the Middle East "must be tipped in Israel's favor."[3] President Johnson received editorial support from some quarters for his decision to withhold sale of the Phantoms, the *New York Times* saying that the president had "struck a remarkably restrained and even-handed policy toward the Middle East since last year's war."[4] Then the Congress stepped in to bring pressure on the president, not that the competing candidates had not already done a good deal of that themselves. In the Foreign Assistance Act of 1968, passed in October, there was included a new section, which read as follows:

SEC. 651. SALE OF SUPERSONIC PLANES TO IS-RAEL. It is the sense of the Congress that the President should take such steps as may be necessary, as soon as practicable after the date of enactment of this section, to negotiate an agreement with the Government of Israel providing for the sale by the United States of such number of supersonic planes as may be necessary to provide Israel with an adequate deterrent force capable of preventing future Arab aggression by offsetting sophisticated weapons received by the Arab States and to replace losses suffered by Israel in the 1967 conflict.[5]

In signing the act, President Johnson added this statement: "I have taken note of section 651 concerning the sale of planes to Israel. In the light of this expression of the sense of the Congress, I am asking the Secretary of State to initiate negotiations with the Government of Israel and to report back to me."[6]

Shortly thereafter the presidential election took place, resulting in selection of Richard Nixon to be the next president, and he was soon propelled into the midst of the Phantom jet decision. Israeli Defense Minister Moshe Dayan came to New York to meet with the president-elect. Only the day before William Scranton, who had been sent by Mr. Nixon on a fact-finding tour of the Middle East, had returned and recommended that the United States adopt a more even-handed policy toward nations of the region. General Dayan commented: "I am sorry you don't have better relationships with the Arab countries and the Soviet Union has so much influence with most of them. . . . I would like the United

States to have more influence in Egypt and Iraq and Syria."
But, he added, not at Israel's expense.[7] Here was encapsulated one of the central policy issues that was to occupy
the attention of the Nixon administration throughout the
various phases of its attempts to restructure the situation in
the Middle East so as to advance U.S. strategic interests and
increase the chances for a negotiated peace. While Israel
wanted to see U.S. influence over its enemies increase, it did
not want to give up any part of the "special relationship" in
order to help achieve that, thereby creating a near-insoluble
dilemma.

A few days later adjacent headlines in a Washington press
summary further illustrated the ambivalence the United
States felt in its relations with Israel. "Johnson Aide Says
Israel Disrupts Peace Efforts" was the heading on an account in the *New York Times* of Walt Rostow's condemna-
tion of an Israeli attack against Beirut International
Airport. Next to it was this heading on a front-page story
from the *Baltimore Sun:* "Sale of U.S. Jets to Israel Reaches
Final Agreement."[8] The latter piece dealt with the government's announcement that Israel would be allowed to buy
fifty F-4 Phantoms, with delivery beginning by late 1969.
"Approval of the incoming Nixon administration," the press
account held, "is implicit in campaign statements of the
President-elect advocating the sale." The Israeli attack in
Lebanon had followed by a single day the announcement of
the sale's approval. Press speculation for the next month
centered on whether the new president would overturn the
decision on the Phantoms, but the sale stood, and the first
F-4s began reaching Israel the following September.

One further event occurred before the inauguration that
was to have far-reaching effects on the Nixon arms transfer
policy. On 8 January 1969 the French government, which
had maintained a selective arms embargo in force in the
Middle East since the 1967 war, extended that embargo to
cover all arms and spare parts for Israel, Jordan, Syria, and
Egypt (the United Arab Republic).[9] The practical effect,
since the principal Arab states were at that time looking to

the Soviet Union for their arms needs, was to throw Israel
almost entirely upon the United States as arms supplier.
The French had long filled that role for Israel, but after the
1967 war they had severely restricted their participation.
One factor determining the willingness of the United States
to provide the F-4 aircraft had been French retraction of an
earlier commitment to sell Mirage V jets to the Israelis. Now
the total embargo meant that United States involvement
would necessarily increase or that Israel would be left in a
militarily untenable position. The problem there, of course,
was that this seemed likely to increase the prospects of superpower confrontation in the region, something that the
United States had long sought to avoid by playing a low-key
role. While perhaps a bit strident in tone, being headlined
"French Maliciousness," the *Washington Post* editorial comment on the French embargo highlighted this point, calling
the embargo "another example of President de Gaulle's ut-
ter disregard for the preservation of peace in the tense and
fragile Middle East" and holding that, "as a political response, the French action is irresponsible." It was, the editors concluded, an example of de Gaulle's "maliciousness."[10]
It was also, given the peak of U.S. involvement in Vietnam
then pertaining, not a very encouraging prospect. But, as it
turned out, that arms dependency of Israel upon the United
States was to be an essential element of the Nixon administration's efforts to achieve progress toward peace in the Middle East.

Shortly after Mr. Nixon assumed the presidency, Israeli
Prime Minister Levi Eshkol died, to be succeeded by Golda
Meir, who was to serve as prime minister until June 1974.
Later she published her memoirs, which include a revealing
passage describing her frustration with the unwillingness to
negotiate of the antisocialist bloc in Israel known as Gahal
and then led by Menachem Begin:

> "But we won't have any cease-fire unless we also accept
> some of the less favorable conditions," I tried to explain
> repeatedly to Mr. Begin. "And what's more, we won't

get any arms from America." "What do you mean, we won't get arms?" he used to say. "We'll *demand* them from the Americans." I couldn't get it through to him that although the American commitment to Israel's survival was certainly great, we needed Mr. Nixon and Mr. Rogers much more than they needed us, and Israel's policies couldn't be based entirely on the assumption that American Jewry either would or could force Mr. Nixon to adopt a position against his will or better judgment. But Gahal, intoxicated by its own rhetoric, had convinced itself that all we had to do was go on telling the United States that we wouldn't give in to any pressure whatsoever.[11]

In September 1969 Mrs. Meir came to Washington to meet with President Nixon and urge that the United States provide more arms to Israel. Before her departure on the trip, the prime minister spent weeks working with Defense Minister Dayan and Chief of Staff Bar-Lev on what she called her "shopping list" of the arms that Israel desired, including a specific request for eighty A-4 Skyhawk and twenty-five additional F-4 Phantom jet aircraft. She also planned to ask for low-interest loans of $200 million each year for five years to help pay for the aircraft. Her purpose in making the trip personally, Mrs. Meir recalled, was to convey to President Nixon "the urgency of our situation and of the imbalance in the flow of arms to the Middle East." Following the meeting, she expressed confidence that the United States intended to continue its policy of maintaining the balance of power in the region and added: "As for my shopping list, well, it eventually changed hands, as it was meant to."[12]

Deciding what to do about such requests, how to maintain the military balance in the Middle East, and in fact what constituted a "balance" under the conditions that pertained there were difficult problems for the new administration. Dr. Kissinger had been quoted as holding, at an early point, that under normal conditions, "if these were two opponents with roughly equal strength, you would say you want to bring about a military balance; but a military balance is death for Israel, because a war of attrition means mathematically that Israel will be destroyed. . . . The Israelis have to aim for superiority."[13] The Egyptian commentator Mohamed Heikal had once made the same point in graphic language: "A military or political balance between the 100 million Arabs and the 3 million Israelis cannot be kept for ever."[14] Another commentator, who claimed to know Dr. Kissinger's personal view, has said it was that "Israel must not be allowed to become too strong militarily while the secret negotiations were under way." Thus, according to this account, "Kissinger's idea, from which he never deviated, was that the United States must use its flow of arms to the Israelis as leverage in efforts to persuade them to give up, sooner or later, territories they had conquered in 1967. This issue dominated American-Israeli relations during Nixon's entire presidency."[15]

While it is unclear at this point whether that viewpoint governed from the very beginning or, alternatively, evolved as events demonstrated its applicability, it is undeniable that in due course this became the U.S. position, one that eventually was encapsulated in the shorthand term "withdrawal for guarantees." And while the administration became convinced that Israel's long-term interests could be served only by negotiating a lasting peace with adequate security guarantees while there was still time and opportunity to do so, there were important elements in both the United States and Israel that were distrustful of that approach. This is why the issue dominated the Israeli-American relationship, as it most assuredly did. But that in itself serves to demonstrate the central role played by the Nixon arms transfer policy in efforts to bring about a peaceful settlement in the Middle East.

Meanwhile, the administration had adopted a go-slow policy in terms of additional arms for Israel and was pursuing a number of diplomatic initiatives in an effort to bring a halt to the War of Attrition being waged along the Suez Canal front. Big Four talks on the Middle East, an idea that had been put forth by President de Gaulle, began at the United Nations in New York in March 1969, with the United States, the Soviet Union, the United Kingdom, and France taking part. At the same time the United States was conducting private and unannounced bilateral talks with the Soviets on the question, urging them to cut back on their deliveries of military equipment to the Egyptians. These latter talks went on between Dr. Kissinger and Soviet Ambassador Dobrynin, meeting in Washington.[16] The results of these initiatives could be gauged by President Nixon's remarks before the United Nations General Assembly in September 1969, where he observed that "failing a settlement, an agreement on the limitation of the shipment of arms to the Middle East might help to stabilize the situation. We have indicated to the Soviet Union, without result, our willingness to enter such discussions."[17]

Despite lack of Soviet cooperation in trying to restrict the flow of arms into the Middle East through joint action, the United States decided to do what it could through unilateral restraint. Thus, in March 1970, Secretary of State Rogers announced that the United States had decided to "hold in abeyance" an Israeli request for additional Phantom and Skyhawk aircraft "in an effort to defuse the military situation and create a climate in which settlement efforts might go forward."[18] At a news conference in January 1970 the president had been asked about the recently announced sale of 100 jet aircraft to Libya by the French and how that would affect our decision to sell jets to Israel. Admitting concern about the potential impact of the French action, the president said that "as we look at this situation we will consider the Israeli arms request based on the threats to them from states in the area and we will honor those requests to the extent that we see—that we determine that they need additional arms in order to meet that threat."[19]

Elements in the Congress again took a hand at this point, with seventy-three senators on 1 June sending a letter to Secretary of State Rogers urging the administration to approve the jet sale.[20] But an amendment that was proposed

to the Military Sales Act, and which was remarkably similar
to provisions ultimately enacted in 1976 in providing that
the administration must seek approval of arms sales on a
country-by-country basis rather than receiving blanket authorization, was defeated overwhelmingly in the Senate in
June. Senator Javits had "warned that it could jeopardize
the sale of jets to Israel."[21] The bill already contained a section expressing congressional support for the president's position that arms would be made available and credits
provided to Israel and other friendly states.

Late in June the administration put forward a new peace
proposal for the Middle East, calling for an extended cease-
fire. Logically there was a connection between the deferral
of additional major arms shipments to Israel and this initiative for an agreement. While a truce did go into effect on 7
August 1970, it was a shaky one, marred, as we have noted,
by extensive violations and some evidence of bad faith. By
early September, the U.S. Congress had passed a new military procurement bill, which provided, among other things,
that there would be no dollar or quantitative limitations on
Israeli military purchases on credit terms. This was, said one
observer, "the most open-ended arms-buying program
offered any country by the United States."[22] It is worth not-
ing that, when the restraints on arms transfers to the Middle
East were removed, the impetus came from the Congress,
not the executive branch. Soon thereafter, Israel received
electronic countermeasures equipment and a supply of the
Shrike air-to-ground missile.[23]

In his memoirs President Nixon recalled his decision of
March 1970 to postpone delivery of Phantom jets to Israel,
based on his hope that, "since Israel was already in a strong
military position, I could slow down the arms race without
tipping the fragile military balance in the region." But, he
continued, reflecting the longer-range goal of his policy, "I
also believed that American influence in the Middle East
increasingly depended on our renewing diplomatic relationships with Egypt and Syria, and this decision would help
promote that goal."[24] If there was one central element run-

ning consistently through Mr. Nixon's approach to dealing with the problems of the Middle East, it was the recognition that, while maintaining a military balance might be essential to achieving peace, it could never by itself be sufficient for doing so.

Mrs. Meir visited the United States again in September 1970 and was reassured by the president that "we did not intend to permit the military balance in the Middle East to be disturbed, and we were prepared to work with her in developing a military aid program that would be appropriate for the strategy the Israelis adopted."[25] Shortly thereafter the president proposed to Congress an additional $1 billion in foreign military and economic assistance, including $500 million in arms credits for Israel, which was approved in the closing days of the legislative session.[26] As the year ended, the intensity of hostilities along the Suez seemed to have died down, with one view being that the administration's strategy of limiting arms transfers to Israel had been successful in that, having no assurance that combat losses would be made up, the Israelis had acted to curtail their deep-penetration raids over Egyptian territory.[27] In his foreign policy report at the end of the year, Secretary Rogers cited the buildup of arms on the West Bank of the Suez, heavy deployment of surface-to-air missiles in the Canal area by the Egyptians with Soviet assistance, and violations of the standstill agreement as causes of concern, saying that the funds that had been requested for Israel would "help restore the military balance."[28]

Two events that were to have major and lasting significance for the course of Middle East relations occurred in 1971, both diffused and extended at the time but sharp in their eventual impact. The first was political, the second economic; negotiations and oil were the issues. In Egypt Anwar Sadat, having succeeded the late President Nasser the preceding September, launched major initiatives toward a negotiated settlement. Extending the cease-fire agreement, he called for renewed Big Four efforts to arrange an interim accord and subsequently stated that he was willing to negotiate a final peace settlement if the interim accord proved a success. The particular significance of the latter step was that it marked the first time that Egypt had signified acceptance of the idea of eventual peace with Israel.[29] But Israel could not be induced to agree to terms. As Joseph Sisco had said, the Middle East had a history of lost opportunities, and surely this was one, for there was no better chance for peace until after war had again swept the region, threatening in the process to precipitate a superpower confrontation.

As for the economic event, it is best described by Edward Krapels:

> In 1971 control over oil began to shift to the producers and brought in its wake a fundamental political realignment within the Middle East. For the first time ever, the lost Arab oil could not be replaced, and the Arab oil weapon became a credible threat. With Nasser's death relations betwen Egypt and Saudi Arabia improved dramatically, and a new political alignment took shape; Israel's occupation of Arab land was its cement, and oil and oil money its sources of international power. This remarkable political change, facilitated by the change in the oil market, confronted the United States with a dilemma. For the first time her support of Israel presented a clear threat to the economic security of American allies.[30]

Were that same assessment to be made today, it is doubtful that the projected security threat could be confined to America's allies, and even in 1973 the impact on the United States itself was sufficient to cause many to think in those terms. Thus, during 1971, in addition to there being no progress toward peace in the Middle East, the portents for the future became, if anything, more ominous.

In late 1971 and early 1972, after more than a year of relative restraint, the United States agreed to provide more military shipments to Israel, including additional F-4 and other aircraft.[31] In February Israel agreed to participate in indirect negotiations with Egypt over the question of reopening the Suez Canal, and the linkage was quite apparent. There was, nevertheless, a good bit of frustration within elements of the administration at this point that withholding of military transfers had not produced more movement on the Israeli part toward negotiations, and there was also resistance on the part of some to renewing deliveries under the circumstances. An additional factor leading to less than unanimous enthusiasm for these particular arms shipments was that the F-4 aircraft to be provided were to be diverted from the production line output scheduled for the U.S. Air Force. As the eighty-six other F-4s provided to Israel over the course of three years had also come from production diverted from the Air Force, some thirty-six of which had not yet been made up, the unhappiness within American military circles was such that the Joint Chiefs of Staff openly pointed out the disadvantages of these shipments and their adverse impact on modernization of U.S. units. The administration position was clearly that it was a price worth paying if progress could be stimulated toward a peaceful settlement in the region.[32]

Reflecting on three years of frustrating and unsuccessful efforts to find a basis for peace in the Middle East, the president's foreign policy report of February 1972 reviewed a series of initiatives that began with "a search for a formula for a comprehensive political solution," involving the Big Four talks and bilateral United States–USSR discussions, then the cease-fire initiative, subsequent efforts to deal with violations of it and the outbreak of hostilities in Jordan, ill-fated efforts at mediation by UN Ambassador Jarring, and finally attempts to achieve an interim solution by concentrating on negotiations over reopening of the Suez Canal. The frustration was obvious in the president's summary remark: "Throughout all these negotiations, each side has sought to influence the other's negotiating position by increasing its own military strength. I have stated on several occasions in the past year that an arms balance is essential to stability but that military equilibrium alone cannot produce peace."[33]

Golda Meir visited the United States again in early March 1973, meeting with President Nixon, who agreed to her request that the United States provide Israel with an additional forty-eight Phantom jets and thirty-six A-4 Skyhawks over the course of the next four years.[34] Later Mohamed Heikal was to write of the effect the announcement of this sale had on the Egyptians: "It was clear from the size of the deal that the United States was once again underwriting Israel's offensive capacity. There was great disillusionment on the Egyptian side."[35] The administration itself was not of one mind on this agreement, and not only because of the impact on U.S. force modernization that was causing the military forces such distress. As was to be the invariable case, the administration was coming under relentless pressure from elements in the Congress to give the Israelis whatever they wanted. Perhaps this, along with the vicissitudes of the last stages of arranging for U.S. disengagement from Vietnam, accounts for the observation Mr. Nixon included in his memoirs, reflecting his outlook as he prepared to embark upon his second term: "I was a man of the Congress and I was proud of the fact. But by 1973 I had concluded that Congress had become cumbersome, undisciplined, isolationist, fiscally irresponsible, overly vulnerable to pressures from organized minorities, and too dominated by the media."[36]

In late September 1973 there occurred an episode that dramatized the many crosscurrents and deep-seated antagonisms that made progress in the Middle East so difficult to attain. Arab terrorists broke into a train carrying refugees from the Soviet Union as it crossed into Austria. Kidnapping seven Russian Jews, the gunmen held them hostage, then used their captives as a lever to force the Austrian government to close down the reception center at Schönau where Jews headed for Israel were processed.[37] Golda Meir had prepared a speech for presentation at just that time to the Council of Europe, meeting in Strasbourg. In it she concluded by quoting European statesman Jean Monnet: "Peace depends not only on treaties and promises. It depends essentially on the creation of conditions which, if they do not change the nature of men, at least guide their behavior toward each other in a peaceful direction."[38] In the event Mrs. Meir discarded her prepared text to make an ad hoc appeal for resistance to the demands of terrorists, but she felt it was important enough to include in her memoirs. Indeed, it had as its essence what the Nixon administration had set from the beginning as its goal, "international relationships that will provide the framework of a durable peace."[39] But before peace could be achieved, there was first to be more war.

The details of the 1973 war in the Middle East are by now well known to all who have an interest in them. Egyptian attacks across the Suez Canal, helped by the achievement of surprise, made major progress on the Sinai front, especially given the outstanding tactical employment the Egyptians made of the antiaircraft missile batteries they had been furnished by the Soviets. A second front was opened by Syria in the Golan Heights region. Three fairly distinct phases of the fighting could be discerned, with the initial Arab successes being followed by a growing stabilization on both fronts and finally a period in which Israeli forces crossed to the West Bank of the Suez and achieved military objectives, including the encirclement of the Egyptian Third Army. From our standpoint, the role played by arms transfer policy in shaping these events, and particularly their aftermath, and the accompanying diplomatic moves are of greatest interest.

Hostilities began 6 October 1973, and from the outset they were characterized by unprecedentedly high rates of attrition in combat equipment on both sides. This led to urgent requirements for replacement of major items of equipment, as well as expendables, most notably ammunition. Mrs. Meir later recalled, however, that "we needed arms desperately, and, in the beginning they were slow in coming."[40] The reasons were not hard to discern. First there was the desire to hold down the flow of arms into the region and the hope that the Soviets might do the same, so that a wait-and-see attitude based on this consideration had some support. Next there was the already substantial body of opinion that the readiness of U.S. forces had been adversely affected by diversions of military equipment to foreign arms transfers, particularly those to Israel, and that further drawdowns should be approached with caution. This viewpoint was to be overwhelmed by the exigencies of the moment, but it was not going to go away and would in fact be greatly intensified by the events of the coming weeks and months. And finally, probably of greatest significance by far, there was apprehension that too forthcoming a reaction in support of the Israelis would trigger an oil embargo.

Presumably the Soviets were engaged in similar calculations, although driven by a somewhat different set of imperatives. We have seen the gratitude expressed to the Soviet ambassador by President Sadat in the wake of the initial Egyptian successes and upon hearing on the evening of 8 October that a Soviet resupply airlift of arms would begin shortly. William Quandt has since hypothesized advance preparation for this support on the part of the Soviets, observing that "some of the equipment captured by the Israelis came from Warsaw Pact stocks, and must have been moved by rail from Eastern Europe to Odessa well before the outbreak of the war."[41] President Nixon apparently "learned from the daily CIA summary [of 10 October 1973] that the Soviet Union had initiated an airlift of military equipment to Cairo and Damascus."[42] As early as 8 October the Israeli ambassador to the United States, Simcha Dinitz, had begun relaying requests for arms shipments to Dr. Kissinger.[43] The following day, the president recalls, he "met with Kissinger and told him to let the Israelis know that we would replace all their losses, and asked him to work out the logistics for doing so."[44] Working out those logistics was to take a bit of doing, and perceptions differ as to just what was in fact going on. That of Mohamed Heikal is interesting as a place to begin:

> The Israelis had organized an airlift from the first moment of the war, diverting as many of their own civilian planes as they could to the United States where they

were enabled, with the express sanction of President
Nixon, to load up with arms and ammunition. It was
when the Egyptians and Syrians heard of this Israeli
airlift that they increased their pressure on the Russians to organize a corresponding airlift for themselves.
Once this Russian airlift to the Arabs had started, the
Americans stepped up their airlift to Israel, making
open use of American planes for the purpose. There
was, however, a conflict between the State Department
and the Pentagon, with some fearing that if the United
States rushed to Israel all the arms that were being
asked for (and the Israeli demands for, above all, antitank guns and missiles had become urgent and almost
hysterical) there would be an adverse reaction on the
Arab side and even the implementation of the much
feared "oil weapon."[45]

Aircraft belonging to the Israeli airline El Al had in fact
begun an airlift from the United States as early as 8 October.
When it was apparent that this was going to prove insuffi-
cient, an attempt was made to arrange for U.S. civilian cargo
aircraft to supplement the airlift, but U.S. carriers were
reluctant to participate in making deliveries to a zone of
active combat. Finally U.S. military aircraft began making
deliveries in large quantities. Obviously, these successive
solutions represent escalating degrees of United States involvement, each stage more overt than the one before it.
Whether this resulted from the understandable reluctance
to trigger an Arab reaction, especially the use of the oil
weapon, or was simply the result of bureaucratic handling
of the situation, I have been unable to determine. In any
event, by 12 October the president "concluded that any further delay was unacceptable and decided we must use U.S.
military aircraft if that was what was necessary to get our
supplies through to Israel."[46] By the next afternoon, he recalled, thirty C-130 transports were on the way to Israel, and
by 16 October a thousand tons a day were arriving there, an
operation bigger than the Berlin airlift (and thousands of

miles longer).[47] The Kalbs quote Henry Kissinger as later saying: "We tried to talk in the first week. When that didn't work, we said, fine, we'll start pouring in equipment until we create a new reality."[48]

The new reality began to emerge by 14 October, when the first flight of C-5 Galaxy aircraft arrived in Israel on the ninth day of the war. "The airlift was invaluable," recalled Mrs. Meir. "It not only lifted our spirits, but also served to make the American position clear to the Soviet Union, and it undoubtedly served to make our victory possible."[49] By 16 October Soviet Premier Kosygin was in Cairo for three days of urgent talks with the Egyptians, and by 20 October Dr. Kissinger was in Moscow in response to an equally urgent invitation from the Soviets. Two days later a cease-fire went into effect, 22 October also marking the completion of one month on the job as secretary of state for Mr. Kissinger. Although the cease-fire fell apart and was subsequently patched up again two days later, the new reality was at hand. It had, however, two faces. The one, brought about by American pressure on the Israelis after the tide of battle had turned in their favor in the wake of the arms resupply operation, was termination of hostilities on the basis of neither side having clearly emerged as victorious or defeated. At one point the threat of a slowdown in the resupply operation had even been directly linked to insistence that Mrs. Meir allow a convoy with food and medical supplies to reach the city of Suez, which had been cut off by Israeli forces.[50] The president had felt that "despite the great skepticism of the Israeli hawks ... only a battlefield stalemate would provide the foundation on which fruitful negotiations might begin."[51] One of those to whom the president undoubtedly referred was Israeli Defense Minister Moshe Dayan, who later told an interviewer: "The U.S. moved in and denied us the fruits of the victory. It was an ultimatum—nothing short of it."[52] The resupply of arms had had its effect on the Arab side as well, such that on 20 October President Sadat sent a message to President Assad of Syria, his most important cobelligerent, to this effect:

We have fought Israel to the fifteenth day. In the first
four days Israel was alone, so we were able to expose
her position on both fronts. On their admission the
enemy have lost eight hundred tanks and two hundred
planes. But during the last ten days I have, on the
Egyptian front, been fighting the United States as well,
through the arms it is sending. To put it bluntly, I
cannot fight the United States or accept the responsibility before history for the destruction of our armed
forces for a second time. I have therefore informed the
Soviet Union that I am prepared to accept a ceasefire
on existing positions.[53]

There was another face to the new reality: On 17 October the
Arab oil-producing states had announced an oil embargo of
the West.

Assessments of the Nixon/Kissinger strategy of trying to
terminate the war on a basis that would leave neither side
in clear superiority have varied among commentators in the
light of subsequent events. Journalist Tad Szulc, in his
lengthy critique of the Nixon era, was harsh in his criticism
of Dr. Kissinger for maneuvering the parties to the war into
a stalemate so that the United States could play the role of
broker in the aftermath. But, as Professor David Calleo observed in his review of the Szulc book, "many readers may
today be more inclined to sympathize with Kissinger than
with Szulc."[54] As to whether Israel was fairly treated by the
United States during the course of the conflict, Mrs. Meir
met with the Knesset on 16 October to tell it about the
success of the Suez crossing and because, as she later said,
she "wanted to make public our gratitude to the president
and the people of America, and, equally clear, our rage at
those governments, notably the French and British, that had
chosen to impose an embargo on the shipments of arms to us
when we were fighting for our very lives."[55] About the president himself, the Israeli prime minister was very clear:
"However history judges Richard Nixon—and it is probable
that the verdict will be very harsh—it must also be put on

the record forever that he did not break a single one of the promises he made to us."[56]

The role played by arms transfers in the continuing unfolding of events in the Middle East did not, of course, end with the termination of immediate hostilities. The president had, while the fighting was still in progress on 19 October, sent a message to the Congress in which he asked for emergency military assistance to Israel amounting to $2.2 billion. There had been formed in the Department of Defense a task force to facilitate the shipment of arms to Israel in what became at times a chaotic and frantic operation. One of its members was later quoted as saying that in the effort to meet Israeli requests "depots were ordered searched, then action came at the next meeting. Israel would ask for, say, a hundred and twenty tanks, and we had to determine the stocks, approve maybe thirty of them, get them moving, then approve more. A lot had to come from army units because there wasn't enough equipment in supply."[57] Continuation of this effort, long after the war was ended, was without doubt the most internally controversial aspect of the Nixon arms transfer policy. While most who were involved conceded that what was done was legitimately in support of U.S. foreign policy, opinions varied as to whether it was worth the risk. Key decisions as to the actual number of major end items to be furnished were made in the Washington Special Actions Group of the National Security Council. This led to some later controversy as to whether the actual Department of State approval required by statute had been provided in advance for every shipment to Israel, but this was a high-stakes game and the decisions were being made at a correspondingly high level.

The equipment provided to Israel in the first month or so included F-4 Phantom and A-4 Skyhawk aircraft, C-130 cargo aircraft, M-60 medium tanks, armored personnel carriers, and large numbers of TOW antitank missiles. More than half a billion dollars' worth of material was shipped in the first few weeks. It appeared to many that the spigot had just been turned on full force and left to run, especially when equipment was taken from war reserve stocks in Germany and from the hands of troops. Even in the first few weeks there was concern in the Department of State as well as in Defense, and efforts were made to get the Israelis to use captured equipment, to enhance their repair capabilities so they could put back into action the damaged equipment they retained, and to hold down additional commitments of actual arms pending a careful analysis of their combat losses, all with a view to minimizing the impact of the resupply operation on the readiness of U.S. forces.

Meanwhile, there was the matter of the oil embargo. While it remains a moot point as to whether the announcement of the major resupply effort for Israel, and particularly the president's request for $2.2 billion to finance arms for Israel, had anything to do with the decision to impose an embargo, there is reason to believe that use of oil as a part of the overall Arab offensive was planned and inevitable. There was the Khalil Report, previously described, which envisioned use of oil in a strategic way. President Nixon later remembered that "the first distant rumble of a possible Arab oil embargo began in the spring of 1973. By mid-summer King Faisal of Saudi Arabia was warning that unless our policy toward Israel changed, there would be a reduction of the oil sent to us."[58] Libya's President Qadhafi was quoted shortly after the onset of the embargo as saying, "we are determined to affect America by striking at Europe [in use of the oil weapon]."[59] In any event, as was later observed, "in 1973 the oil market conducted the shock of the Arab-Israeli-American conflict throughout the world."[60] Following strenuous efforts by the president and the secretary of state, and arrangements for disengagement of Israeli and Egyptian forces, the oil embargo was lifted in March 1974, as we have noted. An Israeli-Syrian disengagement agreement followed in May.

As he neared the end of his truncated presidency, Mr. Nixon was concerned that the opportunity brought about by this complex series of events not be lost. Thus he met with a group of the leaders of the American Jewish community in early June 1974 to express his apprehension over what he saw as their "shortsighted outlook" on what was in the best interests of Israel. He later quoted his journal entry on the meeting:

> I pointed out that hardware alone to Israel was a policy that made sense maybe five years ago but did not make sense today, and that they had to have in mind that each new war would be more and more costly because their neighbors would learn to fight, and there were more of them. . . .
>
> As a matter of fact, whether Israel can survive over a long period of time with a hundred million Arabs around them I think is really questionable. The only long-term hope lies in reaching some kind of settlement now while they can operate from a position of strength, and while we are having such apparent success in weaning the Arabs away from the Soviets and into more responsible paths.[61]

Shortly thereafter the president made his final trip abroad while in office, visiting countries in the Middle East. When he returned, he briefed congressional leaders, telling them that "we would make Israel strong enough that they would not fear to negotiate, but not so strong that they felt they had no need to negotiate."[62]

Secretary Kissinger, for his part, continued to seek ways to advance a long-term agreement. Golda Meir later said that his "efforts on behalf of peace in the area can only be termed superhuman."[63] In tandem with the negotiatory process were additional shipments of arms to Israel. The Middle East Task Group, its meetings now generally reduced to a once-weekly Wednesday session as the ongoing program became routine, tried to ride herd on the transfer requests and monitor the Arab-Israeli arms balance. The Israelis had submitted a plan known as MATMON B, which laid out their projected defense requirements for the ten-year period 1974–1983. A National Security Council study was convened to look at the plan, including its implications for the Middle

East Balance, the readiness posture of U.S. forces, and U.S. defense production. There continued to be strong concern within the departments involved because the Israelis made a practice of bringing those arms requests they knew to be sensitive, either because they affected the needs of U.S. forces or because they involved the introduction of new or significantly more capable military systems into the region, directly to very senior officials of the U.S. government, bypassing the careful review process involved in the normal processing of requests. Contrary to the view subsequently expressed by some critics of the arms transfer policy, there was demonstrated in that process substantial conservatism, especially when it came to questions of proliferating sophisticated equipment, the security implications of releasing advanced technology, and the impact on U.S. force readiness of providing items in short supply. The TOW missile was cited as an example of what can happen. The system had been provided to Israel during the latter stages of the war, and then to Jordan; as a result the United States was "swamped with requests" from other nations who wanted the system.

The General Accounting Office was also looking into the effects on U.S. force readiness of the transfer of arms to Israel and Vietnam. Its report, which was classified, was not released even in unclassified summary until two years later, when it was revealed that during the period 1972–1974 some $8.5 billion worth of military equipment had been shipped to the two countries and that this "adversely affected overall U.S. readiness." Among the details made public were that the Army had withdrawn 369 M-60 tanks and 500 armored personnel carriers from war reserve stocks in Europe and that 400 air-to-surface missiles, representing 5 percent of total U.S. inventory of the item, had been withdrawn from stocks and shipped to Israel during the war. As a result of these and other actions, the GAO found, the Army had a shortage at the end of FY1974 of 4,943 tanks and 1,822 armored personnel carriers.[64]

In September 1974 the Israeli prime minister, Yitzhak Rabin, who had succeeded Mrs. Meir in June, visited the United States. His meeting with Secretary Kissinger lasted

for more than five hours and was followed by press reports
that he "was given assurances . . . of more American military hardware to match recent Soviet arms shipments to
Syria."[65] Former Secretary of Defense Laird, meanwhile,
having left office and associated himself with the American
Enterprise Institute, began to issue statements deploring
provision of "excessive" amounts of arms to foreign customers.[66] While Dr. Kissinger was himself not making public
pronouncements except those calculated to advance the slow
pace of negotiations in the Middle East, a later account said
to have been prepared with the cooperation of some of his
assistants quoted him as saying of the announcement during
the war of the $2.2 billion arms package for Israel, "I made
a mistake," and as going on to say that he considered it his
only big mistake throughout the war.[67] Later, during a difficult period in the negotiations of March 1975, Secretary
Kissinger said that the immense shipments of arms to Israel
without first extracting concessions (again, according to the
same account) "was naive—my biggest mistake."[68]

In the autumn of 1977 defense analyst Anthony Cordesman published a controversial assessment of the provision of
arms to Israel, asserting that U.S. aid to Israel had been such
as to build that country up into "a state able to wage aggressive war with minimal risk." Thus, he went on, the United
States "may now find itself aiding an Israel which may use
its military strength to take permanent control of former
Arab territory in direct opposition to U.S. policy, and be
locked into an indefinite cold war with the Arabs." Recalling
the MATMON B request for more U.S. aid to Israel, Mr.
Cordesman, who occupied a succession of posts in the office
of the secretary of defense during the period in question,
asserted that the United States never accepted the Israeli
view of the threat or the resultant requirements for Israeli
forces. The Israeli requests, he concluded, while scaled down,
were still "incredible." He faulted the United States for failing to take into account in its estimates of the military balance in the area such qualitative factors, all favoring the
Israelis in his view, as training, readiness, command and

control, logistics, defensive barriers, and intangibles.[69] There are many in the Department of Defense who would agree with this assessment, and probably quite a few in the Department of State who would also concur. The late chairman of the Joint Chiefs of Staff, General George S. Brown, who sometimes said impolitic things in public, took a complementary tack in 1976 when he asserted that "Israel had become a burden militarily on the United States."[70]

Meanwhile, Dr. Kissinger continued his extraordinary efforts to negotiate a peace. Golda Meir had expressed particular bitterness in the period immediately after the war toward Menachem Begin and his associates for "talking about a near catastrophe."[71] She resisted the idea advanced by some of organizing an all-encompassing government, which would involve those of Begin's ilk: "I don't want the cabinet to be burdened by an element that would refuse to negotiate —if and when the time comes—because of its totally negative attitude toward any territorial compromise, especially as far as the West Bank is concerned," she said. Now she was out of office, and fortunately her successor was not totally opposed to negotiation. After a spring of frustration during which Secretary Kissinger announced that the United States was once more "reappraising her entire Middle Eastern policy,"[72] an agreement ending this phase of the negotiations was finally reached on 1 September 1975.[73] Press accounts of a concurrently concluded memorandum of agreement between the United States and Israel indicated that it included these provisions:

> The United States government will make every effort to be fully responsive, within the limits of its resources and congressional authorization and appropriation, on an on-going and long-term basis to Israel's military equipment and other defense requirements, to its energy requirements and to its economic needs.
>
> Israel's long term military supply needs from the United States shall be the subject of periodic consultations between representatives of the U.S. and Israeli

defense establishments, with agreement reached on specific items to be included in a separate U.S.–Israeli memorandum. To this end, a joint study by military experts will be undertaken within three weeks. In conducting this study, which will include Israel's 1976 needs, the United States will view Israel's requests sympathetically, including its request for advanced and sophisticated weapons. . . .

The United States government agrees with Israel that the next agreement with Egypt should be a final peace agreement.[74]

Then, according to a later account, there was a "Secret Addendum on Arms Assistance," which read in its entirety as follows:

The United States is resolved to continue to maintain Israel's defensive strength through the supply of advanced types of equipment, such as the F-16 aircraft. The United States government agrees to an early meeting to undertake a joint study of high technology and sophisticated items, including the Pershing ground-to-ground missiles with conventional warheads, with the view to giving a positive response. The U.S. administration will submit annually for approval by the U.S. Congress a request for military and economic assistance in order to help meet Israel's economic and military needs.[75]

The preceding summer, in anticipation of a possible agreement, the point had been broached with the Joint Chiefs of Staff that the Israelis would probably insist on large quantities of weapons as compensation for any negotiating concessions they might make. The Army, in particular, responded with an expression of strong concern about the likely impact on U.S. force readiness. Not only was the U.S. production base uncomfortably narrow in certain key areas, but expansion offered no immediate relief in that lead times of major items were such that it would be two or three years after expansion was completed before any real impact would result. While clearly recognizing and supporting the preeminent importance of achieving a peace settlement if that were at all possible, the Army strongly emphasized that provision of any major items of military equipment that were in short supply and had long lead times for replacement as part of the bargain would inevitably degrade our own readiness posture and urged that this be done only in full awareness of what would result and after careful calculation of what would best serve overall United States interests.

Following completion of the agreement, and in pursuance of the provisions previously cited, a study was indeed undertaken of the expected list of military equipment requirements to be provided by the Israelis. The National Security Council undertook a unilateral look from the standpoint of U.S. interests at the implications of the arms transfers contemplated by the Israelis; included in the factors evaluated were the political and economic impacts, the possibility of compromise of sensitive U.S. technology, and the likely impact of the transfers on the incentives to negotiate of the parties involved. It seems clear on the record that whatever decisions were made were entered into wittingly and after detailed and often agonized (and heated) consideration of the whole range of factors that students of arms transfers are forever citing as relevant. That is not to say, of course, that all parties to the consideration of those factors were happy with the weights assigned or with the resultant decisions. It was in the end a classic case of running a calculated risk.

The agreement of autumn was reached after a spring of frustration. Looking back on that period from several years later, Senator Fulbright remembered a point at which "Mr. Kissinger gave up all his efforts, and said the intransigence of the Israeli government was too great, and that we would have to reassess our policy. Of course, he didn't do that because the Congress would not let him. So we ended up with piecemeal negotiations, which have been of little value."[76] After the breakdown of negotiations in the spring Dr. Kissinger had been quoted as saying that "Israel has no foreign policy . . . only domestic politics." No doubt venting some of his disappointment at seeing another opportunity for an agreement getting away, he described what he termed the peasant mentality of the Israelis: "The peasant is known for his recalcitrance and excessive caution. It is the recalcitrance, the excessive caution, the lack of vision, that have caused the Israelis to refuse this agreement."[77]

Once again the Congress took a direct hand, not in approving the actions of the executive branch but in going a long way toward dictating them. Thus on 21 May 1975 seventy-six senators wrote collectively to President Ford "to endorse Israel's demand for 'defensible' frontiers and massive economic and military assistance":

> Recent events underscore America's need for reliable allies and the desirability of greater participation by the Congress in the formulation of American foreign policy. . . . Given the recent heavy flow of Soviet weaponry to Arab states, it is imperative that we not permit the military balance to shift against Israel. . . . Withholding military equipment from Israel would be dangerous, discouraging accommodation by Israel's neighbors and encouraging a resort to force. Within the next several weeks, the Congress expects to receive your foreign aid requests for fiscal year 1976. We trust that your recommendations will be responsive to Israel's urgent military and economic needs.[78]

A further complicating factor soon appeared. By the summer of 1975 there were public reports that the United States had reason to believe that Israel had acquired a nuclear weapons capability. William Beecher reported that "senior American analysts believe that Israel has made more than 10 nuclear weapons, each with an explosive force comparable to the atomic bombs that destroyed Hiroshima and Nagasaki. The analysts also believe that Israel has the means of delivering such weapons hundreds of miles by plane and missile and is working to extend the reach of its delivery systems."[79] This gave a special urgency to working out solutions to the continuing confrontation in the Middle East and made the idea of constraining the supply of conventional arms an obviously more hazardous proposition than in the past. The prospect of a nuclear-armed Israel with its back against the wall in some future crisis was enough to give pause to all parties to the struggle, no matter how intransigent and demanding it might make the Israelis, who were subsequently publicly accused by General Brown of dictating to the United States Congress. While the general was rebuked for his indiscretion, there were few who sought to contradict him. Anthony Cordesman later quoted Israeli Foreign Minister Dayan (speaking in March 1977, before he assumed that post) as saying, obviously not oblivious to the impact of the image: "How can a country of 3 million people run in this race [the conventional arms race] against the Arabs, who have unlimited financial resources, unlimited political resources for procurement, and huge quantities of manpower? . . . What I am saying is that along with this race we have to develop an option for ourselves, that is, an ability to produce nuclear weapons . . . so that if the Arabs try to conquer and destroy Israel we will have the means to fight them not with conventional, but with high quality arrestive arms—nuclear weapons."[80]

In assessing an administration policy that tries to deal with a situation as difficult as the Middle East military balance, and the appropriate arms transfer policy to prevent an imposed military solution by one of the parties while not inhibiting the will to negotiate of the other, the force and impact of domestic United States lobbies in behalf of foreign nations (some foreign nations, that is) should not be underestimated. We have observed elsewhere the major impact of the Greek-American lobby on negotiations relating to the arms embargo on Turkey. Policy in the Middle East, particularly as it related to continued support for the Israeli position, occasioned even more extensive activity. While a great deal of this was directed at and worked through the Congress, direct public appeals were also common. The means employed were obviously quite well financed and employed attractive propaganda and expensive graphics. A folder distributed as an insert to the *New York Times* provides an example. On the first page, overprinted on an outline map of the region, was shown the "Arab League," stretching from Mauritania and Morocco on the west to Muscat and Oman on the east, and from Syria in the north to Somalia in the south; through a cutout one views Israel in the midst of all this. Opening to the inner page, one finds Israel portrayed all alone on an otherwise blank page, and on the facing page the single question: "Who's threatening who?" Simplistic? Of course. But effective? Undeniably. This particular publication was identified as coming from the Anti-Defamation League of B'nai B'rith, but there were many other organizations devoted to influencing United States policy in this and a number of other issues involving arms transfers. And in an era when senior Israeli officials were talking openly about the need for nuclear weapons, the answer to who is threatening whom is perhaps not so unambiguous as the pamphleteer imagined.

Looking back at these crowded years of the American-Israeli interaction, it is clear that there was a continuing commitment to both the military and the economic well-being of Israel, a determination to make it possible for Israel to negotiate from a position of strength, and a conviction that for its own best interests, as well as those of the United States, it must negotiate a peaceful settlement while that was still possible. Arms transfers were an obvious and even central element in this process, both in maintaining the position of security and in encouraging, indeed bargaining with, the Israelis to negotiate a settlement. Efforts were made by the administration, especially in the latter stages of the period, to moderate the flow of both armaments and military assistance funds, with substantial sentiment within the executive branch for earlier and more sizable reductions than those decided upon. In the event, however, such reductions were rejected by the Congress, which acted to maintain a virtually unlimited flow of both arms and dollars. Whether, as the administration always feared, this would in the long run operate to erode the Israeli motivation to negotiate remains, at this writing, to be fully resolved. While changes in the Israeli government have substantially changed the terms of reference, giving ironic emphasis to some of the earlier misgivings recorded by Golda Meir, the dramatic and unexpected initiatives of Anwar Sadat may prove sufficient, even after his passing, to compensate and thereby salvage yet another chance for peace in the Middle East.

7. Jordan and the Persian Gulf Mosaic

Testifying on behalf of foreign assistance in the summer of 1974, Secretary Kissinger remarked that aid for Jordan was intended to "strengthen Jordan's ability to hold to the course of moderation it has consistently followed."[1] The single comment could serve to reflect both the unique role Jordan has played in the affairs of the Middle East and the reason the United States has consistently sought to support the government of King Hussein and assist it in playing that role. Jordan could, in fact, serve as the very model of what a state with only modest resources, and surrounded by larger and more powerful states, can with courageous leadership do by demonstrating a responsible attitude and a farsighted approach to difficult and emotional issues. Many of the crosscurrents that complicate almost any issue in the Middle East can be found flowing intensely in Jordan. Adjoining Israel, it has had to cope with Palestinian liberation forces and the retributory actions of the Israelis. Abutting Syria, it has been constantly pressured by the more extreme elements of radicalism in the Arab world. Sharing a common border with Iraq, it has had to concern itself with Soviet influence and look to its own defenses. And dwarfed by its Saudi neighbor, it has often been subjected to irresistible political and economic pressures that left it little room to maneuver.

The United States has long served, albeit on and off, as a source of arms for the Jordanian armed forces. Such shipments were halted when war broke out in June 1967 but were begun again the following year in an effort to preserve Western influence in Jordan and to preempt Soviet inroads as arms supplier. King Hussein had earlier spoken with President Nasser, suggesting that he would approach the USSR in an effort to obtain arms, given his disappointment with what he was getting from both the United States and the British. Nasser is said to have advised him to keep trying to get arms from the West, observing that it would not be a good situation to have Israel getting its weapons solely from the West and the Arabs being equipped solely with Soviet arms.[2] A year later, President Nasser told Alexander Shelepin, who had brought him a list of promised arms deliveries, that "King Hussein went many times to see President Johnson to ask him for arms. He never got a single plane. You may be exasperating people to deal with, but in the end you deliver."[3] But Nasser was to change his mind about whether the Russians would in fact deliver, and King Hussein was to get substantial support from the United States. When internal fighting between the government forces and Palestinian guerrillas broke out in Jordan in 1970, that served to stimulate U.S. aid in an effort to keep the king from being overthrown. Both the threat of the guerrilla forces and the intervention of the more radical regime in Syria at that time had caused great concern that the beneficial influence of Jordan in affairs of the region was in jeopardy. Making up the immediate needs of the Jordanians in the midst of the combat situation, the United States switched funds from other country allocations under the security assistance program,[4] a move permissible under the rules in effect at that time but later prohibited by the more stringent and less flexible terms of the annual appropriations legislation.

Lending credence to the view of Washington that President Nasser had attributed to King Hussein, in 1973 a Jordanian party visited the United States to talk about arms, among other things. A key issue involved the TOW antitank missile, on which a commitment had been made to Jordan, with controversy between State and Defense as to the timing of providing the full amount, whether to airlift any portion of the allocation, and other details with more political than military significance. One who was involved at the staff level recalled that Secretary Kissinger sent instructions to be as forthcoming as possible within the restraints imposed by the Congress, so that the Jordanians would leave Washington feeling that we had done our best to help them rather than with the "sour taste in their mouths" they usually had after dealing with the United States.

The Israelis agreed in a general way that it was in their interests for the United States to provide support to King Hussein, and indeed according to some accounts they had been prepared to intervene in his behalf themselves when Syrian tank columns had threatened Amman. In March 1974 King Hussein himself visited the United States and met with President Nixon, who promised additional military aid and also urged the king to do whatever he could to assist in working out a Syrian-Israeli disengagement agreement.[5] A three-year modernization program for Jordanian forces had been developed, including the provision of tanks, armored personnel carriers, artillery, and aircraft by the United States.

Jordan had in particular been seriously concerned about its deficiencies in air defense ever since Syrian aircraft operated freely in its air space during the 1970 fighting. But it was not until the 1974 Arab summit conference at Rabat (at which the Arab leaders, to Hussein's dismay, designated the PLO the sole representative of the Palestinians) that Jordan had the means to acquire air defense equipment, when Saudi Arabia softened the blow by offering Hussein $350 million in subsidies for military equipment. More than two-thirds of that was earmarked by the king for Hawk missile batteries to be purchased from the United States.[6]

Secretary Kissinger had visited Jordan in November 1974. King Hussein raised the issue of his air defense needs, but also laid stress on the political aspects of a major arms acquisition at that time, given his humiliation at Rabat and the adverse impact on his standing with the Arab states externally and with his own armed forces internally. Following the Kissinger visit, the Joint Chiefs of Staff were directed to send an air defense analysis team to Jordan. King Hussein again visited the United States in April 1975, and the Hawk missile arms transfer decision was negotiated at presidential level.

The generally satisfactory arrangements worked out erupted in controversy in the summer of 1975, when the administration informed the Congress that it planned to sell Jordan fourteen batteries of Hawk missiles. In this case the Israelis did not agree, asserting that the proposed arms sale would represent a threat to them, and an intensive lobbying effort ensued. In the face of likely defeat of the proposal through provisions of new law giving Congress veto power over proposed arms sales, the administration withdrew the initial proposal. In a battle that raged for more than half the year, Jordan wound up getting the missiles, but with a number of restrictive provisions as to their configuration and employment, after first refusing to accept conditions, then reconsidering. Because it involved the first use of the Nelson amendment, which came into effect 1 January 1975 and gave the Congress unprecedented authority over individual arms transfer decisions, the case was of importance extending beyond the effects on Jordan or even the Middle East region.

In the midst of seemingly insoluble controversy, the situation was reclaimed dramatically by the conclusion on 1 September 1975 of the peace agreement in the Sinai between Israel and Egypt. Two days later the sale proposal was resubmitted to the Congress, and residual objections were overcome by means of a compromise that specified that the missiles would be installed as permanent fixed systems that could not be used in a mobile role. The president transmitted to the Congress a "letter of assurance" to that effect.[7] Following some further negotiations to induce Jordan to accept the restrictive provisions, the deal was finally concluded in December 1975.

In the last year of the Nixon/Ford administration U.S. technicians were in Jordan to begin installation of the Hawk

batteries, which had finally been approved for sale. While
they seemed certain to improve the air defense capability of
that nation, and hence its resistance to overthrow by outside
forces in any future resumption of the overt military challenge to the regime, the manner in which they were obtained
and the imposition of the restrictions that Jordan had called
"insulting to its national dignity" seemed likely to drain the
deal of whatever political advantage it might have had for
King Hussein. Arms transfer policy was getting more difficult to formulate and implement by the day.

While events along the Arab-Israeli axis received far
more public attention during the Nixon years because of
such arresting developments as active warfare and the possibility of superpower involvement, in geostrategic terms
what evolved in the Persian Gulf was potentially of more
lasting significance. For while the United States had to fight
to stay out of the conflict at one end of the region, at the
other it had an opportunity, and took it, to develop and
support a surrogate willing and able to keep the peace,
thereby making U.S. interests secure without threat of military involvement.

Iran was of course the centerpiece of United States policy
in the Persian Gulf. Again, arms transfer policy was an
essential ingredient of that policy. Citing as examples Iran
and Brazil, an analyst surveying the arms trade had observed that "a new class of states . . . is emerging with the
military and economic capacity to influence events well
beyond its members' borders. As their military power grows
and their horizons expand, they may increasingly behave
like superpowers, perhaps policing affairs throughout a region."[8] This is precisely the role that Iran, with active encouragement from the United States, sought for itself after
the British departure from east of Suez removed that stabilizing influence of long standing. It was not, perhaps, for the
United States the strategy of choice, but given the domestic
political constraints on firsthand foreign involvement and
on an active role for the United States in world affairs, it was
a workable alternative.

The British proclamation of intent to withdraw from the

Gulf by 1971 came in 1968, virtually coinciding with the election of Richard Nixon to the presidency. The vacuum that was created, the increasing importance of oil from the region to the United States and still more to its allies in Europe and Japan, concerns about revolutionary activity in the region and the effect it might have on Western interests, and perhaps most important the necessity to block any Soviet attempts to establish dominance over the region all served to reinforce the tenets of the Nixon Doctrine, with its emphasis on assistance rather than firsthand involvement and the obvious desire on the part of Americans to have some help in carrying the burdens of international responsibility. That the shah was willing to accept a major share of that responsibility was both heartening and gratifying.

As always, it was necessary that there be a conjunction of events and trends to make possible the solution that evolved. Almost simultaneous to the British withdrawal was the beginning of realizing large profits from oil revenues, thereby enabling Iran to move from an economic dependence on outside assistance to an independent and then very quickly to a highly affluent situation. Besides this wherewithal there was in place a competent government and a head of state who not only was strong and desirous of taking responsibility but had demonstrated farsightedness and sophistication in leading his country and utilizing the newfound riches that so fortuitously befell it. Furthermore, the shah of Iran was free of many of the long-standing antagonisms that so complicated solution to the Arab-Israeli problem and was able to deal with both sides of that confrontation without difficulty. As early as 1968, for example, there were reports that Israeli pilots were training in Iran to fly the F-4 Phantom, this during the extended period of negotiations as to whether to provide that aircraft to Israel.[9]

Being strategically placed to interdict the sea-lanes over which the Persian Gulf oil moved to Western markets, Iran was well situated to bring to bear the military power it had acquired and had planned for future acquisition. In addition, it was itself one of the more significant suppliers of oil. According to a contemporary analysis, it supplied 80 percent of the oil imported by Japan and 75 percent of Israel's imports.[10] But the shah's concern to take over responsibility for regional security was far from unrelated to his perceptions of Iran's long-term interests. Basically his approach had been to build up the armed forces so as to protect the oil, then use oil revenues to build a modern industrialized nation rapidly enough to make it self-sustaining by the time the oil ran out. It was a bold strategy and one obviously requiring great skill, energy, and luck to bring off.

It also quite obviously involved a role for Iran whose successful conclusion was in the best interests of the oil-consuming nations. It was far better from the United States' standpoint, for example, to have Iran use American arms to provide stability and guarantee the security of the oil upon which it and its allies had become so dependent than to have that stability break down and then try to go in and do something about restoring the situation, even assuming it would be possible to do so.[11] It is not too strong to suggest that in the literal meaning of the word Iran performed an essential service for the United States and its allies over the decade ending in the fall of the shah's government. And it would be difficult to identify another country outside the U.S. alliance system that was of greater strategic importance to United States interests.

None of this was lost on the Nixon administration. United States support for Iran of course went back a good bit further than the advent of the new administration. In 1959 the United States and Iran had signed a bilateral executive agreement under the terms of which the United States agreed to assist in helping Iran resist aggression.[12] In subsequent years the United States had provided both economic and military assistance, but as the oil prosperity lessened Iran's need for outside help, first the economic assistance was ended (in 1967) and then the military aid (in 1969). As U.S. arms sales to Iran began in modest amounts, so too did the Soviet Union sell military equipment to the shah, with the first shipments arriving in 1968. Press accounts of developments in Iran at the time the Nixon government began concentrated on the shah's wise use of oil revenues. "The reason for Iran's stability and growth has been the Shah's judicious and energetic use of the nation's huge oil revenues," observed one correspondent in a representative analysis, adding that "money has been plowed back into the country with astonishing results."[13]

In October 1969 the shah visited the United States, where he and President Nixon discussed the security problems being created by the upcoming withdrawal of British forces. A study done for the National Security Council in advance of the shah's visit had concluded that "Iran, militarily and economically supported by the United States, would fill the vacuum left by the British. [President] Nixon told the shah that the United States hoped Iran would become the dominant power in the strategic Persian Gulf."[14] As it turned out, not much was required in the way of economic assistance to Iran, and providing the military assistance required was itself to the economic as well as the military advantage of the United States. Given the difficulties the United States was experiencing in meeting the demands of its other allies in various parts of the world, an ally like Iran, which asked for little except what it was willing to pay for in cash and which sought responsibility in roles that contributed directly to U.S. strategic and economic interests, must have looked like a gift from the gods. It is not surprising that some years later, as he subsequently recalled, the president wound up the long day after his renomination "sitting out at the pool and smoking a cigar from Iran."[15]

Iran took diplomatic as well as economic and military steps to build its constructive role in the Gulf. In 1970, after a United Nations mission ascertained that the people of Bahrain preferred independence, Iran renounced its claim to the territory. In his foreign policy report for the year Secretary of State Rogers said that "the United States welcomed this statesmanlike gesture on the part of the Shah."[16] In a summary judgment, the secretary commented that, "stable and progressive, Iran is a constructive force in the region."[17] The cooperation of the United States with Iran in supplying some of its military requirements was also remarked upon. This particular case, given that Iran was to become before the end of the period the largest purchaser of arms and other military goods and services from the United States, provides a useful insight into the deficiencies of dollar volume alone as a measure of anything meaningful about the arms trade. A major factor in the increase in dollar value of arms transfers was the replacement of Britain, with its extensive arms manufacturing capability, by Iran, with virtually no such industrial base, as the principal guarantor of regional stability and security. Necessarily, Iran imported the military wherewithal to fill this role. But the significance is not a correlate of the increase in arms transfer costs; arms were there when Britain was deployed in the region, and they were there when Iran took on the role. In terms of objective reality, some other measure than dollar volume of arms transfers must be found to determine what changes in stability and the prospects for peace, if any, accompanied the transfer of responsibility.

In May 1972 President Nixon visited Iran for consultations with the shah. This was, as it turned out, an historic visit, for the president made the decision to provide the Iranian armed forces virtually any equipment they decided they needed.[18] While this was a decision of the greatest importance in the developing ability of Iran to fill the regional role that both it and the United States wished it to play, the decision also caused unhappiness within the Nixon administration because it had been made at the highest level without preliminary staff-level consideration. Some felt that later individual arms transfer decisions were therefore predetermined, and that this was not a wise approach to what could be a changing situation This was a view endorsed by a later staff study done for the Senate Foreign Relations Committee. And while it is believed that then Secretary of Defense Laird supported the decision at the time, he later expressed misgivings about arms transfers to the region from his consultant's post at the American Enterprise Institute.

The arms exports provided to Iran were both extensive and sophisticated. They included the latest jet fighter aircraft, "smart" bombs and missiles, and other systems that represented the first line of advanced military weaponry. And, because it was later asserted by some that the shah was really interested only in acquiring weapons so he could keep down domestic insurgents, it is important to know something of the nature and disposition of the force he built. It was appropriately and unequivocally tailored to the mission of regional security he had agreed to undertake. Military bases were constructed adjacent to the Strait of Hormuz and on the Gulf. A squadron of hovercraft and marines were stationed on an island in the Gulf near a major refining center. Jet aircraft were dispersed at strategic locations. And naval batteries were installed covering key ocean passages. So far from preparing to repress his people was the shah in his arms acquisitions that a press account at the height of later disorders could be headlined "Lack of Riot-Control Forces Is a Major Iranian Weakness."[19]

Discounting the existence of any outside threat to the Persian Gulf and its oil—"the likelihood of a direct Soviet threat to Gulf nations is today sufficiently remote for that factor to be largely ruled out in the short term as a reason for Western military or naval involvement in the Gulf itself, or for the supply of substantial arms to local states"—Senator Kennedy in 1975 proposed a six-month moratorium on arms sales to Persian Gulf States.[20] The shah's reply went right to the point: "It is not realistic because the Persian Gulf states are not fighting each other. The only fighting in the area has been with subversive revolutionary forces intent on overthrowing the Western-oriented, anti-Communist rulers. It is certainly not in America's interests to deny the Persian Gulf states weapons."[21] Describing the reasons why Iran might wish to possess significant military power at that time, a writer for the *Wall Street Journal* observed that "archeologists still are digging up the ruins of cities which sought to use moral force against the likes of Ghenghis Khan or Tamerlane."[22]

Assessing the results to that point of United States assistance to friendly states in the Persian Gulf, Undersecretary of State Joseph Sisco noted in June 1975 that "the success of these countries in achieving a degree of cooperation and in maintaining the tranquillity that has prevailed in recent years is serving broader U.S. interests in world peace and a relaxation of world tensions."[23] These expectations were realized during the Nixon administration, with continuing programs underway that could have been expected to further increase the ability of Iran to serve both regional and global interests by exerting responsible leadership in the Persian Gulf. Subsequently, of course, Iran was rocked by widespread rioting and determined political opposition, which the shah was unable or unwilling to put down and which instead brought down his government and forced him into exile, radically changing the security situation in the region.

Since this analysis has resulted in a judgment that the Nixon arms transfer policy as it applied to Iran was useful and proper, and that it achieved results that were in the long-term interests of the United States, some further discussion is in order lest these later developments—although they occurred after departure of the Nixon/Ford administration from office—be viewed as undercutting the assessment. That discussion seems to me to have three essential elements: a look at what had gone on in Iran under the shah in realms other than that of the military buildup and what that had to do with development of the domestic opposition; consideration of whether this was something new; and an effort to determine what was different that precipitated the downfall of the shah's regime.

As to the first, it might be useful to begin by repeating Joseph Churba's point about "the sectarian and ethnic nature of Middle East society. Such fragmented societies," he has pointed out, "lacking internal cohesion, are inherently unstable and therefore never fully reliable for strategic sustenance." Thus, he implies, only a strong leadership capable of imposing order can hold such a society together for long.[24] The shah had been such a ruler. He had also not been content merely to preserve the status quo. On the contrary, perceiving the momentary boon represented by the oil revenues, he had sought to transubstantiate the oil into a modern industrialized society in a generation. As one result: "The malaise that many Iranians feel stems from the dislocations that have come to their traditional way of life by the very rapid economic growth for which the shah, and Iran's oil wells, can take credit. Tehran in 10 years has swelled from a city of 2 million to one of 4½ million, and most of the newcomers miss much of the simplicity of their former village life. This is why millions of Iranians are arrayed against modernism and strongly influenced by the Muslim religious leaders' call for a return to a purer more traditional Islamic state."[25]

There is no question, virtually all observers agree, that Iran had made very great progress both economically and socially under the shah's rule. In addition to a vast program of industrialization, per capita income had quadrupled since the 1960s, the literacy rate had doubled, and health care had been greatly improved.[26] According to another estimate, per capita income, which in 1978 stood at $2,000 a year, was ten times what it had been a decade before, and average life expectancy had risen from thirty-five to fifty-two years. Land reform, emancipation of women, and the introduction of profit sharing for workers were other initiatives introduced by the shah.[27] While he had been drastically changing the socioeconomic character of his country, Iranian imports, *not including* military purchases, "rose from $10.3 billion in 1975 to $12.7 billion in 1976, $13.8 billion in 1977 and an anticipated $15.3 billion [in 1978]. That money bought machinery; iron, steel and cement for construction programs; foodstuffs; and so-called turnkey factories, plus the training for their workers and management."[28]

Other indicators serve to underscore the dramatic reforms wrought by the shah. Land cultivated by its proprietors increased from 26 percent of the total in 1960 to 78 percent by 1972. The number of students at all levels went from 1.8 million in 1960 to about 9 million in 1977. Annual rates of growth in the GNP over the decade of the 1970s had consistently been in double figures in constant dollar terms and in 1973 and 1974 reached levels of 50.6 percent and 40.4 percent, respectively.[29]

The budget for the Iranian year running from March 1978 to March 1979 is instructive in terms of priorities and governmental concerns. Defense expenditures, amounting to 700 billion rials, accounted for 17 percent of the total state budget. Development and operation of electric capacity consumed 312 billion rials, education 301 billion, transportation 238 billion, social welfare payments 162 billion, health 101 billion, housing 48 billion, water resources 75 billion, support for industrial development 138 billion, and telecommunications 41 billion.[30] The three principal concerns reflected are national security, improvement in the quality of life for the individual, and development of a self-sustaining industrial base.

In a 1978 collection of essays, *Iran under the Pahlavis,* dedicated to the people of Iran, "whose ancient and unique civilization is experiencing in the mid-twentieth century a spectacular resurgence and progress for the benefit of their own and the world at large,"[31] an international team of scholars with long familiarity with Iran documented the reality of the achievement. By any fair-minded standards, the record is impressive in political, economic, and humanitarian terms. From education to land reform, from public health to the status of women, and from food supply to development of an industrial base, including attainment of one of the highest growth rates anywhere in the world, there was solid achievement across the spectrum. Transportation, reform of the legal system, development of a mercantile middle class, long-range planning, the civil service, national security are all among the aspects of public life that flourished under the shah's rule. The reforms he carried out were characterized by these scholars as being both "more comprehensive and more concerned with social justice and the welfare of the masses" than those of an earlier day upon which he built.[32] The cumulative impact for Iran was characterized as "transition from weakness to strength, from backwardness to progress, and from poverty to wealth."[33]

Opposition that developed to the shah was not all of one political stripe, but the factions were apparently united by one thing, their opposition to the shah's policies. "Both the far left and the far right reject the Shah's liberalization campaign under which he has freed more than 1,100 political prisoners, removed press censorship, cracked down on corrupt officials, permitted formation of political parties and ordered free elections held next June," it was reported in late 1978.[34] The leader of the opposition, then speaking from his seat of exile in Paris, charged that "such liberal policies as emancipation of women and land reform—including nationalization of church land—run counter to teachings of the Moslem holy book, the Koran." An office worker in Tehran, listening to this kind of protest, responded: "It's like a bad dream. These people are going to take this country back to the seventh century." The shah found himself in the middle, opposed on the right "by zealously religious Shiite Moslems who insist that modernization is destroying Iran's traditional way of life and demand creation of an Islamic state" and on the left by "dedicated Communists seeking to oust him and install a Marxist government." Thus, "after nearly four decades on the throne, he is confronted with the real danger of being done in by his driving resolve to modernize his nation."[35]

One might well object that this was a development that had been in progress for an extended period, which is of course quite right. In order further to understand what may have been different in the deteriorating situation that led to the shah's downfall, some extracts from an administration policy document may be helpful. Thus:

The principal objective of U.S. policy toward Iran is to keep this country out of the hands of the Soviet Union. This policy is based upon tangible U.S. interests involved here: the protection of the flanks of the NATO alliance and our Pakistani ally, the safeguarding of the oil supplies to the United Kingdom and Western Europe, the maintenance of a barrier to Soviet political penetration of the Arab world and Africa, and the inhi-

bition to the expansion of Soviet influence and power
which would accompany Soviet access to the Persian
Gulf.[36]

Accompanying that statement of "Our Current Policy toward Iran" was a briefing paper for the president, which
observed that the shah was "suddenly and drastically setting
forth on a program of broad social reform" and that

> The Shah's reform program should in the long run
> bring added stability and prosperity to Iran. It coin-
> cides in large measure with what we have long been
> urging on the Shah [elsewhere the paper recalls that
> the United States had been trying to convince Iran that
> "the greatest threat to Iranian security" was "the in-
> sufficiency of economic development and internal re-
> forms"] and he considers that his course was approved
> in the President's February message of congratula-
> tions.
>
> But for the short run it has involved the Shah in
> additional risks. He has aroused the animosity of the
> dispossessed elite and the fanatical clergy, and having
> not yet consolidated the support of the emancipated
> peasantry, he is dependent in the immediate future to
> a greater degree than ever on the support of the mili-
> tary and security forces. Although these forces might
> not be able to put down a coordinated country-wide
> rising of liberal and urban elements, such a develop-
> ment is unlikely. It is believed that the support of the
> military and security forces can probably deal with any
> internal security problems likely to arise.[37]

Finally, there was the observation that

> The monarchy, which provides the stability not yet
> available through popular institutions or long popular
> experience in political affairs, is in fact the sole element
> in the country that can at present give continuity to

public policy. The Shah, therefore, remains a linchpin for the safeguarding of our basic security interests in Iran.[38]

All of that is interesting enough, but it is even more interesting when one learns that the president for whom it was prepared was John Kennedy and that these are extracts from the reply to National Security Action Memorandum 228 dated 1963, which has only recently been declassified in a sanitized version. The document is entitled "U.S. Strategy for Iran," and as these excerpts make clear, much of it has a familiar ring. Of particular importance, from the standpoint of our present interest, are the strong urging on the part of the United States that the shah undertake drastic reforms, his determination to do so, the recognition that there were risks in his doing so, and the importance of the shah's role to the protection of U.S. interests. From these stemmed the necessity for the United States to provide unequivocal political, military, and (when it was still required) economic support. And this the United States did, for a period of nearly fifteen years, a period during which the shah was able, as we have seen, to bring about remarkable progress in his country. While opposition existed to the social and economic reforms he instituted, it was not able to mount any kind of viable challenge to the shah's leadership. Again we must ask ourselves what changed.

An insight is, I believe, provided by Tom Braden's quotation (in a different, but related, context—the hounding from office of Lyndon Johnson) of an observation by Tocqueville: "Patiently endured so long as it seemed beyond redress, a grievance comes to appear intolerable once the possibility of removing it crosses men's minds."[39] The grievances of the shah's opponents were of long standing. Until recently, the possibility of removing them had not occurred. If we can believe the explanations of those who have now not only glimpsed but seized upon that possibility, we have not far to look for the cause. "Explaining the demonstrations, Medhi

Barzegan, an opposition leader, said, 'When you see a little light, you can't stand the darkness any more.' The opposition credits President Carter's human rights campaign with the light that has been shed."[40] Some confirmation was provided by editorial comments in the *New York Times.* "Support [for the shah] from the Carter administration and Britain's Labor Government has rightly been conditioned on continued liberalization," that newspaper held in late 1978.[41] Intelligence-officer-turned-columnist Cord Meyer underscored the point in a commentary that same week: "The price to this country of President Carter's evangelism on human rights, which prodded the shah into concessions that have put his survival at stake, looks as if it may come high in Iran."[42] The turning point in the fate of Iran may well have been the wire service photograph printed all over the world showing President Carter and the shah of Iran on the south lawn of the White House, eyes streaming from the tear gas police had to use in restraining mobs of demonstrators opposing the shah. The message was clear that the United States could not or would not prevent such activity on the part of "Iranian students," which had to be interpreted by the shah's opponents and observers as at least tacit support for their position.

Dr. Kissinger's views on the rapidly deteriorating situation in Iran were sought by an interviewer near the close of 1978. He was very clear in his assessment: "The Iranian situation is a tragedy for the West. The Shah is a leader who on every critical foreign-policy issue has been totally on the side of the West and who has been a stabilizing factor in every crisis in the area." Pressed for his view as to whether the United States were to blame for the turmoil in Iran, he observed that the process of development tends to produce turmoil, but also cited aspects of U.S. responsibility: "the human-rights campaign, as now conducted, is a weapon aimed primarily at allies and tends to undermine their domestic structures."[43] In another interview published subsequently, Dr. Kissinger contrasted vocal and nonvocal approaches to promotion of human rights in terms of their likely results. "Current developments in Iran are one of the results of . . . a vocal policy," he observed.[44]

Meanwhile it was reported from Egypt, where the shah first went upon going into exile, that he "directly blames President Carter for the collapse of Iran."[45] Other sources belatedly came to the same conclusion. "President Carter's human-rights drive lifted the hopes of the Iranian people to the point of violent protest," reported the *New York Times Magazine.*[46] Far from being grateful for such help, the bene-ficiary—the Ayatollah Khomeini—was reported to have re-ferred to Mr. Carter as "the vilest man on earth."[47] In an even more ominous statement, he celebrated his return from exile by saying, "I beg the Almighty to cut off the hands of foreigners."[48] Further portents of things to come appeared in Khomeini's assertion that it was a "religious obligation" for Iranians to demonstrate against the Bakhtiar govern-ment left in place by the shah upon his departure.[49] By the time Americans in Iran had been seized as hostages, it was clear that we had been misled by Princeton professor Rich-ard Falk, who only a few months earlier had assured us in the pages of the *New York Times* that the depiction of the Ayatollah Khomeini "as fanatical, reactionary and the bearer of crude prejudices seems certainly and happily false."[50] There was even doubt that Andrew Young, then U.S. ambassador to the United Nations, had been correct in predicting that Khomeini would eventually be hailed as a saint (although Young cannily did not specify when or by whom he would be so regarded). For Khomeini had made clear his attitude toward domestic affairs—" If you do not obey, you will be annihilated"[51]—as well as foreign affairs, pledging to Yasir Arafat that when Iran had consolidated its strength it would "turn to the issue of victory over Israel."[52] By way of preparation, perhaps, Iran cut off the supply of oil to Israel. Meanwhile, Abolhassan Bani-Sadr, then (briefly) latest in the series of those directing foreign policy under Khomeini, told employees of the Foreign Ministry, "we are at war and all should cooperate."[53] It all made for very strident counterpoint to President Carter's press conference remarks of a few months before: "I don't know of anything we could have done to prevent the very complicated social and religious and political interrelationships from occurring in Iran in the change of government. And we'll just have to make the best of the change."[54]

Future developments in Iran seem likely to place the shah, particularly by contrast, in an even more favorable light. As events in Iran leading to the departure of the shah were unfolding, television broadcasts of the complete canon of Shakespearean drama began in the United States. The opening work was *Julius Caesar,* which contained a line that seemed particularly timely. Caesar having been done in by the conspirators, a citizen is heard to confide his foreboding: "I fear there will a worse come in his place." In like manner it is not necessary to portray the shah as without fault in order to appreciate his appeal in contrast to the evolving alternatives.

I believe the assessment of the Nixon arms transfer policy with respect to Iran and the foreign policy that it served in the Persian Gulf holds up even in light of more recent events, even especially in light of them. Those policies were undermined by changes the Nixon administration would not have countenanced, but that does not depreciate the value of the original policies. Had those policies been continued, there would have been no need for British Foreign Secretary David Owen, referring to the Iranian government of the shah, to ask, as he did: "Can you just take their money, sell them tanks for strategic interests, sell them cars, persuade them to hold down oil prices in the interests of the world, generally exert influence with them and then, when they come under attack, just back off?"[55] For as the *Washington Post,* commenting editorially on "The New Iran," reminded us, "if Americans are to be honest with themselves, they cannot ignore that for 25 years they respected the shah for his modernization policies and profited from his cooperation in the economic and strategic spheres. Let us not pretend now that we didn't know the bad as well as the good—that he ruled harshly and had strategic fantasies, but was a useful and reliable ally who could be counted on to advance American interests, indefinitely. Successive administrations bet heavily on him, getting more from him perhaps than they had any right to expect."[56] Nor would there have been the need to try, as the Carter administration was reported attempting to do at the end of 1979, to build up Egypt in military capability beyond its own defense needs "in hopes of adding stability to the Middle East."[57] That stabilizing influence existed in Iran under the shah. Failure to exert strenuous efforts to maintain it can only be viewed, in geostrategic as well as economic terms, as an irretrievable blunder of immense proportions. And, sadly, it was also in any meaningful terms—whether of liberty or of prosperity—a major and possibly permanent setback for the people of Iran. As events continued to unfold, the United States and Iran were in an undeclared state of war, and few dared deny that the worst was yet to come.

Along with Iran, the capacity for maintaining the stability of the Persian Gulf region and the security of the oil resources located there came to depend heavily on Saudi Arabia. Perhaps the most conservative regime in all of the Arab states, the Saudi royal family has been determinedly anticommunist, antirevolutionary, and devoted to business. With a population far smaller than Iran's, a large and rugged expanse of territory, and bountiful oil resources, it has security problems challenging enough for any nation. It is not surprising, then, that during the Nixon years the Saudis became major customers, in fact the premier customers in terms of deliveries, of the United States for military equipment and services. Based on cumulative deliveries for FY1950–FY1976/7T, the Saudis ranked first in total dollar value of military sales, first in dollar value of purchase of weapons and ammunition, and first in dollar value of purchase of weapons alone.[58]

These are the kinds of figures one is accustomed to hearing with respect to Saudi Arabia, but they are far from telling the whole story or even focusing on the most important aspect of it. For the fact is that the Saudis, perhaps reflecting that very conservatism for which they are so noted, invested the vast majority of their military expenditures in construction, training, and other ventures designed to improve their ability to man, maintain, and employ the military hardware they were acquiring. During the 1970s, for example, only 20 percent of the Saudi military expenditures abroad were for weapons, while two-thirds of the total went for construction projects.[59] Thus, while military insecurity has been viewed by some as Saudi Arabia's "one weakness,"[60] the steps it has been taking to eliminate that weakness have been measured and planned. United States policy emphasized the dual reliance upon Iran and Saudi Arabia to maintain regional stability and in particular to prevent the establishment of regional hegemony by any outside power or the undermining from within of their respective regimes. The obvious concern on the part of Saudi Arabia in the face of the downfall of the shah's regime in Iran is further evidence of the stake both felt in maintaining mutual stability and in cooperative action.

Saudi Arabia, like Iran, is a country in which the United States has for a number of years provided not just arms and other military goods and services but also training and technical assistance aimed at building up the indigenous capability of the defense force. As early as 1970 Secretary of State Rogers indicated that in Saudi Arabia "Americans provide technical expertise in a variety of fields, including . . . modernizing Saudi Arabia's armed forces."[61] By 1973 this program had advanced to the point where the United States was willing to provide sophisticated aircraft, air defense systems, and other advanced weaponry to Saudi Arabia, always along with the elaborate infrastructure the Saudis insisted upon. Although, according to one account, a consultation with Dr. Kissinger in the summer of 1973 about whether to accede to a Saudi request to buy F-4 aircraft produced only "snuffling noises," the Saudis got the F-4. Presumably there was a good bit more snuffling, for they eventually got a great deal more.

In addition to their efforts to get more weapons, and more advanced weapons, for themselves, the Israelis complemented those efforts by trying to block American arms sales to other nations in the region. The prospective sale of F-4s to Saudi Arabia was one such transaction, with the Israeli ambassador meeting with Assistant Secretary of State Sisco and with Dr. Kissinger to urge that the sale be disapproved and with a variety of Israeli diplomats talking openly to newsmen and public figures to draw attention to their objections. While the State Department stood by the decision, pointing out its desire to meet the "legitimate security requirements" of countries in the area, the American Israel Public Affairs Committee issued a "we are dismayed" statement concerning the proposed sale.[62] But the snuffle held. Meanwhile, in the Arab press the American decision to sell F-4s was greeted as a significant change of policy, and a welcome one.

In the autumn, of course, the war intervened in these developing relationships, but not as disruptively as might at first have seemed to be the case. King Faisal was, it has been reported, in secret communication with the president and Secretary Kissinger even during the course of the fighting: "He anticipated that Nixon would replenish Israel with arms to compensate for the Soviet deliveries to Egypt and Syria, but he warned the President that a massive, public bequest to Israel would make Saudi forbearance impossible before Arab opinion."[63] Whether in the event it was quite this clear-cut or not, the underlying point that Saudi Arabia was, like other states, not entirely a free agent—that it had to take into account the positions and perceptions of its Arab neighbors, as well as those of subgovernmental elements such as the Palestinians—is obviously both correct and pertinent. Also, if the Egyptian study as to how to utilize oil to best advantage during the war was put forward as suggested by Heikal, the Saudi response may even be judged moderate under the circumstances.

If anything—and this was a point no doubt perceived by the Saudis and other Arab states—the United States' interest in dealing on good terms with those states was intensified with the coming of war. Not only did the regional security concerns continue unabated as a source of common interest, but now the influence of the Saudis in helping to resolve the Arab-Israeli dispute, and in helping to end the oil embargo, became acutely important. Thus Saudi Arabia became something of a special case in terms of U.S. foreign policy interest,

primarily because of the strong influence it exerted, both
politically and economically, over other states in the region.
Political, rather than arms transfer, considerations clearly
took precedence in dealings with Saudi Arabia at this time.
And, eager as the United States was to find opportunities to
demonstrate its evenhandedness in dealing with Arabs and
Israelis, the provision of arms to an Arab state, and one that
could probably be trusted to use them in a responsible way,
was an attractive policy option under the circumstances.
This reinforced the already strong arms transfer connection
existing between the United States and Saudi Arabia, more
than counterbalancing any tendency to disruption of that
connection that the war and the oil embargo might have
been thought to bring about.

These factors serve to emphasize the intensely political
nature of arms transfers, whatever their military significance. In this instance the means of moderating the short-
term effect, so as to minimize heating up the already
agitated Middle East arena, was the present agreement to
future deliveries of weapons systems. In the Saudi case, this
was made easier by that government's predisposition to undertake systematic development of the infrastructure and
preparatory activities before trying to assimilate large quantities of new weapons. Thus the military impact of the arms
transfers was deferred, as it were, while the beneficial political impact was acquired when it was needed most.

Further influencing the decision to continue with the supply relationship were two additional factors. Increasingly,
the Saudis and others were able to acquire arms from European manufacturers, and in fact they had on occasion even
been encouraged to do so. More importantly, the use of the
oil embargo was aimed not at ensuring supplies of American
weapons but at seeking to force American influence to be
brought to bear on the Israelis with a view to recovering for
the Arab states their occupied territories. Arms were of far
less significance to the Arabs than attaining this objective,
and thus there was no possibility that shutting down the
supply of arms would achieve U.S. objectives, whereas doing

so would lessen the chances of being able to get the Saudis to help in that respect.

In 1974 a Saudi-American defense commission was established, and the motivations that had led the United States to serve as arms supplier to the Saudis caused it also to make "a determined effort to retain its position as the dominant supplier of arms to Saudi Arabia against competition from Britain, France and other Western nations."[64] These suppliers had taken advantage of a delay in acting on requests from Arab states in the aftermath of the war, when a large number of applications for export licenses just stacked up as the United States pondered the policy options. While no formal decision had been made, the effect began to be that of a de facto embargo. Eventually the resolution of the oil embargo eased the strain, and things started to move again, while some selective relief had been provided in the case of Saudi Arabia even before that in terms of filling prior commitments. Meanwhile, a delay of six weeks in obtaining State Department approval of an order for water trailers (a great many water trailers, in fact) to the Saudis provided a bit of comic relief, causing one high official to wonder in writing how, if we could not get the water trailers on the way with a bit more dispatch, we were ever going to be able to achieve the "extraordinary cooperation" between the United States and Saudi Arabia that had recently been proclaimed as the policy objective.

Under this increasingly close cooperative arrangement, U.S. military missions were sent to Saudi Arabia to help evaluate the security needs of that country and assist in the planning of force and equipment development responsive to those needs. "It is our view," Undersecretary of State Sisco told the Congress, "that the major burden for assuring security in the region must be borne by the gulf states themselves and in particular by the major nations of the region, Iran and Saudi Arabia."[65] Reacting to suggestions that the United States might do better not to supply so much military equipment to the Saudis and others in the Gulf region, Mr. Sisco responded that "we see no practical way to separate the military and defense aspects of our policies from the diplomatic, political, economic, and other ties we maintain. We cannot claim friendship and interest in one breath and deny goods or services which have life-or-death importance with the next."[66]

The results of this policy, beyond the long-term and comprehensive concern for maintenance of regional security and stability that was served during the period, are that the Saudis have underwritten Egyptian military and economic needs, supplied funds for a military program in North Yemen that brought that country under its influence, and provided the means for Somalia to acquire needed arms elsewhere than from the Soviets. Dominating OPEC, Saudi Arabia blocked a price increase in June 1976 that nearly every other exporter favored. But perhaps most significant of all has been the support provided Egypt, which was literally the means of President Sadat's ability to disengage his country from dependence upon the Soviet Union for arms and from the consequent influence of the USSR on Egyptian affairs and political freedom of action.[67]

Kuwait, along with Saudi Arabia, Bahrain, and Oman, gets all of its arms from the West. Kuwait was declared eligible for foreign military sales by the United States in 1971. Subsequently the United States sent several small military teams, at Kuwaiti request, to help survey defense requirements of that country. In 1973 Kuwait undertook a seven-year program to modernize its armed forces, with the equipment for doing so to come from the United States, France, and Britain.[68] A Kuwaiti team visited the United States in the spring, and letters of offer for various military equipment had been prepared by the United States in advance of the visit. The following month U.S. military representatives were in Kuwait, where they found British and French arms salesmen aggressively competing for the business.[69]

In addition to a generalized threat stemming from Kuwait's extremely large oil reserves and its weak military posture, a more particularized threat involved a territorial dispute with neighboring and much stronger Iraq. Kuwait had been a British protectorate for nearly a century until gaining independence in 1960 and even then had enjoyed British security guarantees until the withdrawal of British forces in 1969. Thus, providing for their own defense was a new experience for the Kuwaitis. In March 1973 a serious border dispute erupted, with Iraqi forces entering territory claimed by Kuwait and killing two border police officers. Only vigorous intervention by other Arab states persuaded the Iraqis to withdraw. There followed a demand from Iraq that Kuwait give up two islands controlling the access channel to a new port on the Persian Gulf that Iraq was developing with Soviet assistance. The islands had been given over to Kuwait as recently as 1963 when Iraq relinquished its claim to them in exchange for a large loan, which incidentally it had never repaid. This series of episodes served to inject considerable urgency into Kuwait's efforts to improve its defense capabilities.[70]

Following the U.S. survey of its defense needs, Kuwait made arrangements to purchase quantities of Hawk air defense batteries, A-4 Skyhawk jet fighter aircraft, and TOW antitank missiles. There were no illusions as to what these augmentations would enable Kuwait to do militarily. An administration spokesman explained that they were intended to provide the means of slowing down any aggressor long enough for Kuwait to be assisted by friendly forces in the region or for diplomatic intercession by other states to take place.[71]

Dealings with Kuwait illustrate some of the conflicting objectives that inevitably had to be reconciled in making decisions about arms transfers. In 1973 the Kuwaitis wanted to buy the F-4 Phantom fighter aircraft, but the United States preferred that they take a less sophisticated system, such as the F-8 or the F-5. This position was based on a general desire to hold down the level of technology going into the area, the belief that if the Kuwaitis were given the F-4 then other states in the region would also want to have it, and the projected impact on American forces of providing such late-model items. On the other hand, the French were trying to sell Kuwait the Mirage, and the United States preferred to keep both the influence and the business. As it turned out, this was one of the cases in which the staff-level recommendation (not to sell the F-4) was overridden at a higher level.

During the course of U.S. efforts to assist the Kuwait armed forces in their modernization efforts, various senior military officials from the United States made visits to Kuwait. One of these was Admiral Thomas Moorer, then chairman of the Joint Chiefs of Staff. Later he told of reading a newspaper in Kuwait that carried an article inveighing against Kentucky Fried Chicken because it was run by that infamous West Point graduate and strategist who planned the Israeli war, Colonel Sanders. Admiral Moorer observed that it made him kind of homesick—he thought he was reading the *Washington Post*. Meanwhile Kuwait, the richest nation in the world in terms of per capita income, continued efforts to develop at least some capacity to protect itself and its riches, at least long enough to provide time for outside intervention in its behalf if nothing else.

One of the last major endeavors of Richard Nixon's presidency was his trip to the Middle East in June 1974. On the eve of his departure he noted in his journal these thoughts: "not just this trip when it is concluded, or the two and a half years remaining when it is concluded, will mean that we have secured our goal of lasting peace. It is going to require tending thereafter by strong Presidents for the balance of this century. And who knows what can happen thereafter."[72] While it seems there is always much more to do in bringing about a genuinely lasting and stable peace in this region, certainly a great deal has been achieved over the course of the decade since the Nixon administration came into office. As the foregoing analysis has sought to demonstrate, purposeful policies on the part of that administration, policies in which arms transfers played an essential and central part, had much to do with what was achieved.

Golda Meir once observed that "the most essential public service for all in Israel, unfortunately, is the army."[73] With the help of the United States, that was made and kept a strong army. Henry Kissinger argued that "we should not support the proposition that only those who are prepared to go to war can determine the conditions of peace."[74] With the help of the United States those who wished for peace were made strong enough to help preserve it. In their biography of Dr. Kissinger the Kalbs concluded that "he seeks only stability. For him, there is no higher form of international morality."[75] While that may well be correct in a limited sense, it is important to understand the distinction between stability on the one hand and maintenance of the status quo on the other. Stability does not imply the absence of change. Indeed, the negotiations the Nixon administration fostered in the Middle East were designed to bring about profound change. But the imperative for stability is that change be orderly, that it encompass the legitimate interests of all the parties involved, and thus that it not sow the seeds of future discord and violence by inducing a set of conditions that cannot be tolerated over the longer run. While it may seem paradoxical to some that introducing arms, sometimes in large quantities and in the most sophisticated versions, can contribute meaningfully to stability, thus serving the cause of both peace and justice, that is precisely what was demonstrated by the Nixon arms transfer policy as it was carried out in the Middle East.

8. NATO and West European Arms Transfers

The realm of arms transfers to West European nations, and particularly to those forming part of the NATO alliance, is very much a special case when it comes to evaluating the Nixon administration's arms transfer policy. Only with respect to these transfers were the critics relatively uninvolved. These were developed nations, undeniably able to allocate a portion of their budgets to defense expenditures, faced with an observable threat in the East, possessing the trained manpower and technical sophistication to operate and maintain modern weaponry, already clearly linked to the United States by an overt mutual security agreement, and—some argued—paying too little as their share of the costs of defending Western Europe.

Thus, sales to these nations were not susceptible to many of the objections raised with regard to other regions of the world. There was also the need to offset balance of payments problems, which the United States was experiencing as a consequence of stationing in Europe its military forces contributing to the alliance. And of course there was the element of competition between U.S. defense manufacturers and those of European nations, principally France and the United Kingdom, which were the most important alternative Western suppliers of sophisticated weapons systems. In this case, therefore, U.S. firms sold their hardest and were in consequence extremely successful. So much was this the case that frequently voiced alliance aspirations to make the business of military sales more of a two-way street, with U.S. purchases from West European manufacturers of defense equipment increasing to provide more of an equilibrium, never really came to fruition. Late in the period, largely as a component of the "arms deal of the century" involving competition to provide the next generation of fighter aircraft to NATO forces, an extensive series of agreements providing for coproduction and subcontracting arrangements made the first meaningful progress in this direction.

Meanwhile, the pressures described—and their relative absence in the United States-to-Western Europe arms trade—led to a result that may have been the most counterproductive aspect of U.S. arms transfer policy in this administration. Instead of working out agreements that would systematically provide for Western European arms manufacturers to hold a significant and planned share of the market served by Western exports, thereby assuring their continued viability and serving the desirable end of maintaining independent armanents industries in several Western nations, the United States competed so aggressively that it deprived its allies of much of this opportunity. To the extent thay were able nevertheless to garner a sufficient market share to maintain viable domestic arms industries, this occurred in spite of—rather than because of—United States policy. While the result was thus advantageous, even if fortuitously so, in military and economic terms, it was decidedly not in political terms. The European resentment of American insensitivity to their legitimate needs in this realm was a net political loss that no doubt affected other aspects of the relationship as well.

The reality was that, while every nation aspired to an independent capacity to manufacture the arms it required for national security, only the superpowers could do so without extensive exports to spread the costs of developing and producing a full range of defense materiel. The second-tier nations, while not able in any case to develop on their own the entire range of advanced equipment that characterized the armed forces of major powers, could nevertheless approach that if they could sustain an appreciable export trade. Britain and France asserted that, for them, this meant exporting half of what they produced in their armaments industries.[1]

The other side of this coin, of course, is that few nations can hope to provide for their own military needs entirely, or even substantially, from domestic production. Thus purchases of arms from producing states are of critical importance to their security and ability to play a stabilizing role in their regions. This aspect has been discussed extensively elsewhere in this study: the point at the present juncture is that it appears decidedly disadvantageous for the United States to pursue a policy that could, if carried to its logical conclusion, result in its becoming the sole Western source of major items of defense equipment, either to other members of the NATO alliance or to arms recipients elsewhere. While the point could be pursued in considerable detail, perhaps it is sufficient to point out the undesirable impact on the strategic balance of a situation in which the poststrike environment would contain no alternative sources of conventional armaments available to the West, even had a nuclear exchange involved only the superpowers. While the importance of this is clearly linked primarily to the deterrent aspect of the force balance, poststrike recovery capabilities are widely acknowledged as being an important element in such calculations. The preservation of those capabilities inherent in the armaments industries of other Western powers would thus appear to be very much in the interest of the United States, as well as of the alliance and its other members—thus my contention that United States policy might better have provided for a sustaining share of the export market in arms for certain Western allies.

The United States is not the only arms-producing nation that has delayed deliveries to its own forces, or even withdrawn equipment from them, in order to meet the needs of its arms customers. It has been reported, for example, that in the fairly recent past "France's allies and many of her own military experts [were] viewing the state of French military forces with increasing alarm. Their fears stem from definite indications that French conventional forces . . . are being further crippled in the name of the country's aggressive drive for arms exports."[2] This serves as an indication of how much importance France attaches to exports as a means of maintaining its arms industry and the resultant independence of reliance on foreign suppliers: "If political problems or delivery dates have presented difficulties in making a sale, France has been more than willing to have her armed forces suffer the resulting consequences. When delivery dates proved an impediment to closing deals to sell Mirages to Spain and Dassault-Breguet Atlantic patrol bombers to the Netherlands, France captured both contracts by withdrawing similar planes from French squadrons and lending them to those countries until the Spanish and Dutch could be sent models fresh from production lines," the report continued.[3] Similarly, French forces have waited to receive new antitank and air defense missiles of French design while overseas customers have been given priority on delivery.

The problem of maintaining some self-sufficiency in the face of brutal market pressures is also exemplified by recent developments in the arms industry in France. Aerospatiale, the government-owned military conglomerate that manufactures aircraft, missiles, and space equipment, is the result of successive mergers of what were once six separate companies formed in the 1930s. Consolidation and cooperative ventures with other European companies have been necessary to survive. An overwhelmingly important element in that survival is production for the export market. Following a long string of unprofitable years, the corporation began making money in its missile, helicopter, and space divisions, with only aircraft continuing to show a loss. But it exports three-quarters of its helicopter production (which includes versions for civil as well as military use). Sixty percent of the production of its missile division is for export. The success of its portable Milan missile and the larger HOT missile and of the Roland antiaircraft missile built in partnership with West Germany has given Aerospatiale a strong export position, enabling it to compete successfully (profitably) with firms having much larger domestic markets for their products, notably American manufacturers. The importance, even essentiality, of the export market to maintaining the domestic armaments industry is obvious.[4]

French military forces have also been told to consider export potential specifically in drawing up the specifications for their own military equipment needs. Industrywide, there is a consciousness of the importance of designing to the needs of the export market.[5] In trying to carve out a segment of the market where French products could enjoy a competitive advantage, the focus has in some cases been on more easily maintainable and somewhat less sophisticated items of major equipment than those usually supplied by the United States but still more advanced than what recipient nations could produce for themselves or acquire elsewhere. A respected retired French general officer, Pierre Gallois, had suggested such a strategy in the early 1970s, arguing that the United States had only "second-class" aircraft like the F-5 or very complex and expensive ones like the F-14/F-15 to offer and that there was a middle ground that the French were in an excellent position to dominate. In this particular realm, he asserted further, France should forsake partnership opportunities entered into merely for the sake of sentimental "European cooperation," reserving cooperative development for those cases that both promised significant advantages to France and increased its competitiveness with the United States.[6]

Answering charges that the export thrust was dangerous to the readiness of French armed forces, a French journalist specializing in defense affairs summed it up as follows: "Everybody is pleased with this policy. France keeps hundreds of thousands of people working and is in a position to equip its army under favorable conditions. Customer countries are escaping from superpower military pressure to a certain extent, while buying excellent material at decent prices; and those weapons are generally easier to maintain because they are built with a typical French sense of economy."[7] As in so many other of the cases we have considered, the economic and strategic factors are here complementary, but it must be remembered that the strategic and political factors are dominant and that the economic imperatives are of a parcel with the overall considerations of national security that have driven the French to try to maintain an independent capacity to formulate and prosecute their own national security policy. And, as suggested earlier, there are important advantages to the United States in France's being able to do that. Here the arms transfer policy of the Nixon administration was deficient, in my view, in failing to take sufficient account of this possibility and in not devising means of assisting the European nations to maintain a share of the export market. How far from this position the actual policy was is perhaps best illustrated by a conversation with a senior military officer assigned to the office of the Joint Chiefs of Staff to work on security assistance matters: "We might *want* the French to get the [arms sales] business?" he asked. "I never thought of that."

The British for several years have also been following an active arms export policy and for much the same reason as the French. A Defence Sales Organisation created in the mid-1960s was tasked to "insure, within the limits of government policy, that as much military equipment is sold overseas as possible and also to develop research to stimulate the interests of buyers."[8] It is no wonder that recent United States suggestions for cooperation in restraints on arms exports have been rebuffed by both the British and the French. The *Guardian of London*'s defense correspondent summed up some of the British views on the importance of the arms export market: "Britain and France are both former imperial powers equipped with advanced military technology that their own armed forces can no longer economically sustain. As a result, there is great pressure in both countries to export armaments whose development and production costs would otherwise be prohibitive. The revenue can then be used to support national defense. Britain's arms exports, for example, approximately meet the cost of keeping an army on the Rhine as part of the North Atlantic Treaty Organization."[9]

During the late 1960s and early 1970s, West Germany entered into agreements to purchase substantial quantities of arms from the United States specifically as a means of helping to offset the costs of maintaining U.S. forces in Germany as part of the NATO defense establishment, which deployments were causing a heavy drain on the U.S. balance of payments position. In the last such agreement, which ended at the close of FY1971, the West Germans spent about $1.5 billion over a two-year period. But the continuation of such a means of providing offset became less feasible as German military requirements were more completely met, and after FY1971 the Germans offered other forms of assistance, including providing funds for renovation of such U.S. facilities in Germany as troop barracks.[10]

Another factor contributing to lessened willingness on the part of the Germans to spend large amounts for armaments imported from the United States was the redevelopment of their own indigenous arms manufacturing capacity, which had of course been reduced to nothing following the close of World War II. Not only cooperative ventures with, for example, the French but also the manufacture of major items of defense equipment of German design lessened the need to buy so much from overseas suppliers. This development paralleled a change in arms export policy on the part of the Germans as well. In 1974 the Arms Control and Disarmament Agency report had summed up the German export position as follows:

> Notwithstanding its great industrial potential, the Federal Republic of Germany has limited the development of its arms industry and has pursued a relatively restrictive policy with respect to arms transfers to countries outside of NATO. . . .Exports to non-NATO countries are greatly limited both with regard to destination

and type of equipment. A strong desire to avoid association with conflict situations has led the Germans to adhere closely to a policy of not exporting arms to areas of tension. Economic pressures to export apparently have not been a major factor in German arms transfer policy.[11]

By the late 1970s this policy had undergone substantial change. Naval ship sales led the way, with Indonesia, Ghana, Abu Dhabi, Iran, and Nigeria among the West German customers, as well as nearly all of Latin America.[12] This policy change, it seems clear, was not unrelated to the long and frustrating experience the West Germans had undergone in the last several years with efforts to obtain from the United States the often-promised agreements to purchase substantial amounts of West German military equipment, thereby helping to provide the market volume needed to sustain manufacture at a viable level. More than half the decade was consumed with abortive efforts to codevelop a main battle tank, with the end result coming to little, and that late, of major significance to the Germans.

These developments give rise to a whole series of questions of importance for arms transfer policy: Are we better off having the Germans make these sales, and whatever others they may choose, than making them ourselves? How many of them are sales made in lieu of those the United States declined or was prevented from making? Do we think the Germans will exercise the same, or greater, discretion in making sales decisions? Will they seek to obtain the same, or superior, results from whatever leverage they thus obtain? Are these benefits, if there are any, worth the tradeoffs, both economic and political? Do we want the Germans to develop more independence in arms manufacture, thereby reducing their arms dependence upon the United States? Can we do anything about it? Have we made a mistake by failing to make good on our repeated commitments over an extended period to purchase more weapons ourselves from the Germans, which would have provided a more viable base for their domestic arms production capabilities? And has this been a factor in causing them to revise their policy on exporting arms outside NATO?

The answers to these questions, which typify the difficulties involved in formulating arms transfer policy, even assuming that all or most of the outcomes were within our capacity to control, are far from obvious. But what does seem obvious, and it is the central point to be made with respect to arms transfer policy toward the West Europeans, is that the United States could have achieved its goal of greater standardization within NATO and at the same time provided risk-free outlets for European defense materiel by buying more from its NATO allies, and it could have exercised much more control over what kinds of arms went to what kinds of recipients by entering into an agreement with its West European allies to allocate shares of the export market (it must be emphasized that this is not the same as an agreement to *restrict* exports, which is anathema to the French and British in particular for the reasons cited).

In the mid-1970s a four-nation consortium of NATO members, including Belgium, Denmark, the Netherlands, and Norway, undertook to obtain a common replacement for their aging squadrons of F-104 fighter aircraft. As the competition to meet this requirement shaped up, it came down to three candidate systems: the Saab Viggen Euro-fighter from Sweden, France's Dassault-Breguet Mirage F-1E, and the F-16 built by the General Dynamics Corporation.[13] The initial sale of 348 aircraft, worth some $2 billion, was expected to lead to further sales of whichever system was selected to Spain, Canada, and other nations in Asia, Latin America, and the Middle East, amounting to perhaps as many as 1,500 additional aircraft. Thus the appellation of the "arms deal of the century," which quickly attached to the competition, seemed to fit the facts of the matter.

After a year-long contest marked by high-level political interaction and sometimes publicly expressed bitterness among the competitors, the F-16 was selected. General Dynamics, which had had to win a prior competition with the

Northrop Corporation to field the American entry, was jubilant. A spokesman at the corporation's Fort Worth plant, virtually shut down as the F-111 production line ground to a halt, called it a "once-in-a-lifetime reprieve."[14] And, since the F-16 was clearly quite an airplane, there seemed to be grounds for satisfaction based on military considerations as well. But still there were cautionary voices raised, and along the lines of the disadvantages of U.S. arms transfer policy toward Western Europe we have been discussing. One such caveat was presented editorially by the *New York Times:*

> [The F-16 sale] raises the serious question whether the true long-term interests of either the United States or its allies have been served by this so-called "arms deal of the century."
>
> No objective of American foreign, economic and defense policy in Europe has been more long-lasting or more frequently stated since World War II than that of helping West Europe stand on its own feet—at least, in the defense field, in terms of conventional forces. . . .
>
> The gradual decline of West Europe's military-industrial complex, moreover, has weakened the very forces which everywhere provide the most potent political support for the kind of vigorous defense programs Washington usually urges on our NATO allies. . . .
>
> . . . unless American support is given to the revival of Europe's arms industries for essentially European purposes, Europe will never carry its weight in the common defense of the democratic alliance.[15]

This argument made excellent sense and highlighted once again the ambivalence that had been so pervasive a part of United States policy toward Europe since the war. On the one hand, there was the desire to have an independent—and if possible united—Europe operating as an important center of political, military, and economic power in the world, one that could relieve the United States of some of the burden of carrying the responsibilities for serving the interests of the free world. On the other hand was the lingering reluctance to relinquish the high degree of control America had enjoyed in European affairs since the war. The two inclinations were, of course, incompatible, and policy wavered between the two pulls, and not just in the realm of defense and arms transfers.

Even though the U.S. design had won the competition, Europeans preferred to obtain major weapons systems from their own manufacturers, thereby enhancing their economic well-being and as a result their security, for the two factors are inextricably linked. In fact, as one witness told a congressional committee, European nations had come to feel so strongly about this that they were willing to reduce their purchases of American weapons systems even at the cost of also reducing their military capabilities.[16] This is not as anomalous as it might sound at first, for the point is that the economic viability of their defense industries, and of the larger national economies in turn, are themselves key aspects of the national security and must therefore be given consideration in the context of decisions of defense production just as with the other security determinations.

As more was learned of the terms of the F-16 agreement, it became apparent that these feelings on the part of the European nations that had agreed to purchase the F-16 had been largely accommodated within the terms of that sale. Far from being the last of the straight buys from the United States, it represented much more a complex implementation of the philosophy that indigenous arms industries must be sustained in the interests of national security. In fact, although it was done at European insistence rather than in pursuance of a U.S. policy, at this point the Nixon arms transfer policy might be said to have been reformed in important respects as it applied to Western Europe.

The arrangement called for the F-16 to be produced contemporaneously in the United States and Europe, the first time such a plan had been used for manufacture of a major weapons system. The production-sharing arrangement was a key element in getting the Europeans to agree to buy the F-16 and was in fact an integral part of the bargaining that preceded completion of the deal. As it was worked out, the European partners would build 40 percent of the parts for the aircraft they themselves were to receive, 10 percent of the parts for those being acquired by the United States for its own use, and 15 percent of the parts for F-16s sold to other nations that were not parties to the agreement. There were to be a total of three assembly lines, one each in Belgium, the Netherlands, and the United States.[17] The result was expected to be that, given a projected sale of some 1,500 aircraft total, the participating European nations would recoup between 63 and 100 percent of the cost of their national orders due to their production participation. And when production reached its peak, probably sometime in 1983, an estimated 23,000 workers would be employed at the European manufacturing facilities.[18]

The Atlantic Council had convened a working group on security, cochaired by Harlan Cleveland and Andrew Goodpaster, to produce a report on "the growing dimensions of security," which was published in the autumn of 1977. Emphasizing the economic and political aspects of national security, not instead of but in addition to the military considerations, the report concluded among other things that the United States had been "highly culpable in subordinating NATO needs to national considerations, particularly in the procurement of equipment."[19] The conclusions we have drawn substantiate and reinforce that judgment, with respect not only to sharing the export market and buying military equipment from the Europeans but to selling to them as well. But the F-16 arrangements went a long way toward establishing a model responsive to the latter concern.

Arms were specifically involved as quid pro quo, or combinations of arms and economic assistance were used, to obtain or maintain U.S. base rights in several cases in Europe, including Spain, Greece, and Portugal (in the Azores). The United States also contributed to a multilateral economic package that compensated Malta for allowing British forces assigned to NATO to use bases there and denying base rights to Warsaw Pact forces.[20] Several of these arrangements, and particularly the Spanish case, were the subject of prolonged

controversy between the executive and legislative branches
of the United States government, within the larger context
of efforts to restrain the authority of the executive which
have previously been discussed.

One additional case, however—that of Turkey—is deserv-
ing of a closer look as part of consideration of arms transfer
policy and Europe. Turkey, along with Greece, Taiwan, and
South Korea, was one of those states the United States had
designated "forward defense countries" due to their locations on the periphery of the communist world. As such they
had long been the focus of security assistance. A fairly typical attitude of developing nations toward arms was expressed by Turkish Premier Nihat Erim on a visit to
Washington in 1972. "We want every kind of weapon necessary for a modern army," he said.[21] The American interest
in that aspiration was increased by the fact of Turkey's
membership in the NATO alliance and by the fact that Turkey contributed more than a third of the total land-force
strength to the forces of that alliance.

Not that relations with Turkey were without their vicissitudes for the United States. Turks were not all that happy
about any real or implied dependence upon the United
States or the alliance. "We have been neighbors of Russia for
600 years. We have fought 13 wars with her, and we have
survived without the United States and NATO," one Turk
pointed out to a visitor. And Turks had not, to a man, forgotten that the United States intervened to stop a Turkish
landing on Cyprus during the 1964 crisis there.[22] And it was,
of course, another Cyprus crisis a decade later that propelled
arms transfer policy toward Turkey into the center of domestic political controversy in the United States.

Partly because of this background, partly because of the
timing, Turkey is in many respects the most instructive case
of all. It demonstrates the more encompassing goals of those
who sought to limit arms transfers more severely than the
administration was already doing. It likewise illustrates the
limited leverage provided by arms transfers and the kind of
reverse leverage that accrues to a recipient who has accepted

arms as a quid pro quo for something (base rights, in this case) the United States wanted.

This was a somewhat indeterminate case to begin with, as it was not all that clear that the Turks had in fact violated the provisions of security assistance agreements that held that arms provided could not be used except for defensive purposes. For while the Turks did use the arms to invade the island of Cyprus, something they readily conceded, a case can be made that they did so for defensive purposes, that is, to protect the Turks residing there from harm at the hands of the Greek elements that had been involved in the coup that overthrew the legal government of the island.

What was involved was in reality a very minor skirmish between Turkey and Greece over Cyprus, precipitated by a power play backed by the Greek military junta, to which the Turks responded by moving in military forces. The International Institute for Strategic Studies later observed that "it was symptomatic of the junta's incompetence that it should have believed that Turkey would not react vigorously."[23] In a campaign of active hostilities lasting no more than a month, there resulted at most some few hundred casualties on the two sides combined. It was, by quantitative standards at least, not much of a war. And the sequence of events regarding Turkey, Greece, and Cyprus included overthrow of the antidemocratic military junta in Greece that had started the whole thing, a result that should have delighted many in the United States who had been advocating such an outcome for some time.

Even if fair-minded observers could conclude that the Turks had violated the terms upon which the United States had provided them arms, subsequent open-ended and potentially permanent punishment of them for so doing seems open to question on a number of counts. In the first place, the embargo on arms shipments imposed on Turkey had clearly at some point served the ostensible purpose of penalizing the Turks for violating the rules, as well as demonstrating to other recipients of American arms that they could expect similar treatment if they should violate them (although Israel subsequently did so, in invading southern Lebanon, without penalty). While it was certainly possible to differ as to how long the embargo should last to accomplish this purpose, clearly that point came and went long before the embargo was ended, which did not occur while the Nixon/Ford administration was on the scene.

Thus the only really significant adverse effect of the whole affair was induced by Americans who insisted, over the strenuous and repeated objections of successive presidents, on imposition of a punitive, prolonged, and coercive arms transfer embargo on Turkey. The effect was to put Turkey in an untenable bargaining position with respect to Greece over the Cyprus issue, since resumption of arms transfers was tied to progress in reaching a Cyprus agreement. Greece could afford to hold out without bargaining or making any concessions and in fact could, from its point of view, derive some positive advantage from doing so, since the longer the matter remained unresolved the greater the adverse impact of the embargo on Turkish armed forces, thereby weakening them relative to Greek forces at the same time the pressure mounted on Turkey to capitulate in the negotiations. No Cyprus agreement resulted, of course, as could have been predicted by any serious student of the influence derived from arms embargoes. While providing or withholding arms has substantial and divers uses, like other instruments of policy it invariably fails to influence foreign nations to do or refrain from doing things they view as seriously adverse to their interests as they perceive them.

The pro-Greek lobby in the United States was extremely active in seeking to prolong the arms embargo of Turkey, while any pro-Turkish domestic lobby in the United States appeared to be nonexistent. Thus the mutual reinforcement of those who wished to constrain the administration on any grounds (the Cyprus crisis occurred less than a month before the Watergate denouement and Mr. Nixon's resignation), those who wanted to reduce arms transfers on whatever grounds, and those who wished to undermine Turkey to the supposed benefit of Greece was sufficient to overcome repeated administration attempts to lift or modify the ban.

Perhaps the most dramatic illustration of the more general alienation and desire for withdrawal from American involvement abroad which then pertained was the persistence with which the embargo was clung to in the face of the obvious and increasing damage it did to the military posture of the NATO alliance and the strength of the Turkish adherence to that alliance. For the longer the embargo continued, the worse became the status of Turkish military equipment and the less able their military forces to fulfill their NATO mission. Likewise, the increasingly emotional alienation and political ambivalence engendered by the embargo served to undermine the Turkish attachment to the alliance. The Soviet Union, not oblivious to these developments and anxious to capitalize on them as much as possible, made overtures to the Turks that eventually were not turned aside. Thus Soviet Prime Minister Kosygin visited Turkey, where he opened a $600 million steel mill that had been financed by Soviet loans, and the two countries also agreed to draw up a pledge of mutual friendship and cooperation. Bulent Ecevit, then the leader of the political opposition in Turkey, suggested the idea of a nonaggression pact with the Soviet Union, and press speculation was that if he ever returned to power he would probably build on and accelerate the changes then underway.[24] Subsequently, of course, Mr. Ecevit did return to power, whereupon he stated his view that "for many years Turkey has been carrying too heavy a burden for NATO, allocating to NATO a proportionately greater part of her national income, her budget and her manpower than any other member country." Furthermore, he argued, "historically and geographically Turkey is primarily a Balkan, Middle Eastern and Eastern Mediterranean country. . . . we should give greater emphasis to these historical and geographical realities."[25] Further disadvantage to the United States itself resulted from the retaliatory closing of U.S. bases in Turkey, bases that reportedly served primarily to support intelligence-gathering efforts directed at the Soviet Union.

In the midst of the controversy over the Turkish arms embargo, the *Washington Post,* in a perceptive editorial, went to the heart of the matter. "Conflict with Congress over sending arms to foreign states has become the central problem of the Ford administration's foreign policy," it observed. Further, "it arises from all of the strains of the last few years —indeed, of the last generation—which have weakened not only this President but the presidency in matters of international affairs. Typically, the administration tries to sell or grant military aid to fulfill a 'commitment,' stabilize or strengthen a client, or otherwise serve a larger strategic or diplomatic purpose. And the Congress, either not sharing that purpose or not perceiving it, resists." The editorialist identified as "the nub of the problem" this situation: "Congress refuses to accept, as any President must, the responsibility for creating and executing an overall diplomatic plan. But it reserves for itself a right to torpedo whole enterprises by vetoing particular steps to put them into effect."[26]

"We are concerned," the *Post* explained, "with the effectiveness of the general approach to these issues." It went on to advise the president that, until he could obtain congressional backing for the broad outlines and objectives of his foreign policy, he was going to continue to be frustrated in the details by congressional unwillingness to back, fund, or authorize his programs in support of that policy. But this was exactly the problem: The Congress, for whatever reasons, was having no part of a foreign policy that still envisioned an active role for the United States in world affairs. To the contrary, in the wake of Vietnam, Watergate, and economic problems of inflation and recession and unemployment, and with the convenient excuse of détente in its most utopian prospects, the Congress as a body was simply not interested in providing the wherewithal for continuing U.S. obligations and potential U.S. involvement in overseas conflicts.

The reality was that the president *could not* obtain congressional acceptance of the broad outlines of his foreign and strategic policies, and therefore he had either to alter those policies or to seek to achieve them to the degree he was able despite congressional lack of support, and even opposition.

Arms transfers were one of the key means of doing that, and so they became one of the constant sources of controversy between executive and congressional elements, eventually evolving into, as the editorialist recognized, "the central problem" of the administration's foreign policy. While it did not change things much, no doubt the administration was still gratified by the *Post*'s conclusion that "the President . . . has every right to expect that the Congress will not keep spoiling his strategies with crippling attacks on their specific ingredients. Congress, which so often reacts emotionally and politically on the basis of limited information and partisan insight, is wide open to attack on this score."[27]

In its application to Western Europe the Nixon arms transfer policy, as constrained and inhibited by elements of the Congress and domestic political opinion, thus had two principal difficulties. "Producers provide arms to nonproducers because they deem it useful to do so," the Council on Foreign Relations study had observed.[28] The failure of the United States to be more sensitive to how useful it was to West European manufacturers of arms to do so, and to recognize that it was in United States interests to ensure that they had an adequate and reliable market for doing so—rather than competing with them to such an extent that at one point both France and Britain were moved to enter official complaints that the United States was using political pressure to minimize European export sales success—was a principal flaw in the administration's arms transfer policy.

Henry Kissinger later cautioned that "we must take care not to erode the distinction between allies and neutrals."[29] In the case of Turkey, we failed to observe that sensible guideline. The administration fought desperately under two presidents, however, to prevent the Congress from imposing the arms embargo on Turkey, then sustaining it beyond all possible productivity. It was not a failure of policy that it was unable to make that case. If anything, it was an untimely intersection of foreign policy interests with domestic political vicissitudes reinforced by an uncongenial global outlook.

9. Latin America, Asia, and Africa

Arms transfer policy with regard to Latin America was inherited by the Nixon administration and was legislatively determined rather than being the result of any administration intentions. Imposing rigid constraints, or at any rate attempting to do so, on Latin American arms acquisitions appeared to be done not because there was much of a problem there but because it was easy to do to a group of friendly and relatively weak neighboring states.

Administration policy, as differentiated from that mandated by the Congress, had been consistent and coherent on the matter of arms transfers to Latin America for a number of years. During the Kennedy administration, for example, the Department of State had issued instructions concerning U.S. policy on Latin American military purchases that directed diplomatic posts to press Latin American countries to limit military expenditures and avoid purchasing unnecessary military equipment but to encourage them to make those purchases they did require from the United States. Ambassadors were authorized to use the threat of reduced military and economic assistance to dissuade Latin American governments from unwise military purchases.[1] That guidance was implemented with remarkable effectiveness by United States diplomats and military advisory personnel. Undersecretary of State Nicholas Katzenbach told an audience in 1968 that hemispheric military budgets had declined in real terms by almost half over the course of the last two decades.[2] At the same time Secretary of the Air Force Harold Brown pointed out that Latin American nations as a group spent less than 2 percent of their combined gross national product for military purposes, lower even than in Africa; that Latin America was receiving only 7 percent of the military assistance being provided by the United States to foreign nations; and that the economic aid programs for Latin America supported by the United States were fifteen times as large as those for military aid.[3]

Meanwhile, United States military assistance and arms sales made available to nations of Latin America, while maintained at this level, had provided access to the governments concerned, preempted inroads by potentially hostile nations seeking to use arms transfers to their own purposes, and provided a means of moderating conflicts within the region should they erupt by restricting the flow of replacement systems and spare parts when necessary. Against this background there arose the determination of elements in the Congress to further restrict Latin American nations in their efforts to obtain arms, at least from the United States.

Many nations in Latin America were armed with weapons systems of World War II or Korean War vintage, which had naturally become older and thus more difficult and expensive to maintain, as well as in some cases more dangerous to operate. In 1967 the Congress added a provision to the foreign assistance legislation placing an upper limit of $75 million on military assistance and sales to Latin America, which represented a $10 million reduction from the level of the previous year. There was also introduced the Conte amendment, providing for reduction of economic assistance by the same amount as developing nations spent on sophisticated weapons. Restrictions on particular systems were also introduced, such as those on supersonic jet aircraft.

Peru acquired the first supersonic aircraft in Latin America in December 1967 when it placed an order for twelve Mirage V fighter-bombers with France. Although the Inter-

American Summit Conference held at Punta del Este in 1967 had resulted in a statement of recognition that arms purchases retarded development, the statement of intent to limit military expenditures was sufficiently hedged to result in its having negligible impact.[4] This is not so surprising, however, when one considers that the Latin American military expenditures were themselves, by world standards, close to being of negligible significance. The following spring the United States cut off foreign aid to Peru in retaliation for the jet purchase. Peru thereupon threw out the U.S. military mission. Shortly thereafter it was reported that the State Department "looked with disfavor" on Peru's purchase of six used Canberra bombers from the British but "acknowledged it could do nothing about it."[5] When a projected sale of M-16 rifles by the United States to Brazil took three years to go through the process for issuance of an export license, Brazil canceled the deal.[6] Argentina announced a "Plan Europa," which involved switching over to European sources of weapons, then manufacturing them in Argentina under license, and eventually selling on the export market.[7] Arms dealer Samuel Cummings quoted a high Argentine official as saying, as he placed an order for the French AMX-30 tank, "We are sick of the United States."[8]

Before long, it was reported that Latin American countries were "awash with salesmen of arms and planes from Britain, France and Sweden, all anxious to sell the Latins what they can no longer obtain from the United States either by grant or purchase."[9] Brazil bought Mirage supersonic fighter-bombers from the French; Argentina bought missile-carrying destroyers from the British. It was announced in the press that the Nixon administration, "seeing lucrative orders going in quick succession to France, Britain, West Germany and Italy," had "modified the policy of the Johnson era by declaring its readiness to supply the expensive armaments."[10] But modifying the policy was one thing, and getting the Congress to go along was another. Quoting from his address of the previous autumn on Latin America, the president observed in his first foreign policy report that "we have sometimes imagined that we knew what was best for everyone else and that we could and should make it happen.... Experience has taught us better."[11] It was to become painfully evident in later debates on arms transfers that the experience of which the president spoke had not imparted the same lesson to all. There were many who maintained that the Latin Americans were investing in arms they did not need or that were too complicated for them to handle and that the United States should refuse to sell. There developed a continuing conflict between those who asserted this viewpoint and others who felt that this represented a kind of residual paternalism that was foredoomed and that the United States was unwise to forfeit the advantages of its former position as the dominant military supplier for Latin America in an effort to attain a result that could not be so gained.

Secretary of State Rogers provided some figures to go with the thought: "In the 1966–1970 period, U.S. military sales to Latin America, under congressional limitation, averaged $24 million a year. In the same time lapse, Latin American equipment purchases from 'other nations' reportedly soared to between $500 and $550 million.... U.S. inability to 'meet what they consider legitimate security requirements [has] been interpreted by many Latin Americans as an example of paternalism and lack of concern for the region.'"[12] Administration efforts continued to get the ceiling on arms transfers to Latin America removed or modified, but these met with only modest success. By 1972 the $75 million ceiling for the region had been raised to $100 million. Sponsoring an amendment to further liberalize it, Senator Hugh Scott updated the experience with Latin Americans and their reaction to the unilateral U.S. restraints: "During the combined fiscal years 1970 and 1971, U.S. sales to Latin America, cash and credit, totaled about $130 million. Military material grants to the region added another $14 million. During the same period, Latin American countries have purchased military equipment in the amount of at least $900 million from European suppliers."[13] As other senators rose to support the proposed amendment, some of their colleagues grew restive, presumably due to their own unwillingness to relax the restraints. The record reads at that point: "SEVERAL SENATORS. Vote! Vote! Vote!"[14] But Senator William Brock was allowed to speak, and he stated his view very simply: "it is clear that a number of Latin American nations are deciding to modernize their equipment inventories with the only real question being the source of supply."[15] Secretary Rogers reminded the Congress in his testimony that the Latin Americans were still spending only about 2 percent of their GNP for military purposes and observed that more than a third of what the United States sold to them consisted only of spare parts, follow-on support, and an assortment of minor items.[16] An editorial writer commented on the effects of the restrictions: "Not surprisingly, the Latin Americans, who think they know more about their security problems than Washington, began buying what they wanted from France, Britain and other willing suppliers."[17] One of those willing suppliers, as it turned out, was the Soviet Union, which increased the percentage of its trade with less developed countries represented by Latin America from 0 percent in 1970 to 12 percent by 1977.[18]

In testifying on the proposed FY1974 security assistance program, Deputy Secretary of Defense William Clements made some points that were to sound familiar a few years later:

Current legislative restraints on Foreign Military Sales are self-defeating. This is not to say there should not be restraints on arms sales to Latin America—or any part of the world. The restraint should be exercised by means of case-by-case assessment of purchase requests—based on the criteria of need, ability to pay, effect on neighbors and responsibility for use. This type assessment currently results in refusal of purchase requests that are not consistent with total U.S. interests. Arbitrary ceilings have the effect of putting the countries at odds with each other to obtain U.S. articles and services, of alienating the prospective buyer against the United States Government and forcing him to buy from other sources.[19]

Deputy Secretary of State Kenneth Rush tried again to make what seemed a very simple point to the Senate Foreign Relations Committee: "Recent experience has demonstrated that the Latin American ceiling has not restricted arms spending but has simply diverted it from the United States to Europe. As a region, Latin American nations still spend less than 2 percent of gross national product on their defense budgets. Since we cannot control even this limited spending, we believe that it is to our mutual advantage for Latin American countries to meet their equipment needs through United States sources."[20] Responding to this logic, President Nixon exercised the waiver authority contained in the law to find Argentina, Brazil, Chile, Colombia, and Venezuela eligible to purchase the F-5E figher aircraft.[21]

Finally, by 1974, the administration was able to have the restriction on cash sales to the region lifted and the credit sales ceiling for the region set at $150 million per year. Opponents switched to other tactics, Senator Kennedy introducing an amendment, for example, calling for a ban on arms aid to Chile until it reformed its practices with respect to human rights. Much was made of the convening of representatives of eight Latin American nations in the Andean region to conclude the Declaration of Ayacucho, wherein they expressed their desire to create "conditions which will make possible the effective limitation of armaments and put an end to their acquisition for purposes of war." Three years later Helga Haftendorn reported that "in the meantime all major signatories have not only purchased supersonic aircraft and other advanced weaponry but also accelerated their indigenous arms production."[22] Presumably the desired conditions had not yet been attained, although the opportunities for cutting back are limited when expenditures amount to only 2 percent of GNP to begin with.

Something of the impact was indicated when it was revealed that during the period March 1974 to March 1975 the Department of State's Office of Munitions Control had rejected 253 proposed arms sales.[23] In 1976 the situation was further dramatized when the Soviet Union sold Peru both SU-22 aircraft and a large number of tanks.[24] Meanwhile, during FY1976, U.S. arms sales to Latin America comprised less than 1 percent of its total sales worldwide.[25] Ecuador, concerned about its neighbor Peru's acquisition of Soviet aircraft, sought to purchase Kfir fighters, which incorporate a U.S. engine, from Israel. The United States vetoed the sale. Next door, several hundred Soviet advisors arrived to help operate the Russian tanks and other equipment.

Stimulated by the withholding of American military equipment, Brazil in particular set out to develop an indigenous arms manufacturing capability. As early as 1975 the Brazilian army spent $64 million on domestically produced replacement weapons and equipment, items it would formerly have imported. A Brazilian Industrial Development Commission (CDI) was formed to oversee the work of Imbel (the Brazilian War Materiel Industry); CDI's secretary subsequently confirmed that "arms sales on the foreign market are a major part of the strategy stipulated for Brazil's arms industry."[26] Brazilian arms exports have already had some significant successes, notably in the field of wheeled armored vehicles. Engesa (Engenheiros Especializados SA) of São Paulo builds two models that have already been sold in large numbers to countries in the Middle East, including Libya, Iraq, and the United Arab Emirates, and are in use in the Brazilian army as well.[27] To the south, Argentina has put into production a "Third World" tank built to German design, called the TAM (Tanque Argentino Mediano). Pakistan was said to have ordered 400, the first model rolled off the assembly line in December 1977, and according to a veteran observer Argentina's next step would be to attempt to compete energetically in the export market.[28]

A tabulation of the major publicly known Latin American orders for military equipment during the last three full fiscal years of the Nixon/Ford administration shows France, Britain, the Netherlands, Israel, Italy, Brazil, Canada, West Germany, Australia, and the USSR as suppliers of arms in addition to the United States. Besides 200 T-55 tanks, the Soviets provided artillery and helicopters to Peru. Israel sold surface-to-surface missiles to Argentina and transport aircraft to Mexico and Ecuador. France sold armored cars to Chile and patrol boats to Venezuela, among other clients. West Germany sold submarines, Italy frigates, Canada transport aircraft.[29] Stanley and Pearton concluded that "the efforts of the United States to dam the tide of Latin American arms purchases were a conspicuous failure,"[30] and that was *before* Latin American indigenous arms manufacturing capabilities were stimulated to serious efforts. The tide had really been a trickle, but the insistence upon trying to stop even that had served instead to increase both the flow and its independence.

The successor administration, despite its interest in restricting the traffic in arms, confirmed the relative absence of a problem with respect to Latin America. A State Department fact sheet opened by providing this as background: "Latin American nations traditionally have displayed a cautious attitude toward arms purchases, and tend to give priority to economic development. Most do not feel threatened sufficiently to justify priority for external defense requirements."[31] Lucy Wilson Benson, undersecretary of state for security assistance in the new administration, confirmed this view and the results of past U.S. policy in an address to the Woman's National Democratic Club in Washington, citing "the historical example of Latin America, where we have exercised restraint over the last 10 or 15 years with the result that the Europeans now have 70% of the Latin American arms market."[32] The Council on Foreign Relations study further established the point: "For years the United States refused to sell supersonic combat aircraft to Latin America because there appeared to be no requirement for such high-performance weapons and because all available planes were needed in Vietnam. The policy was unsuccessful—Latin

American countries acquired jets from other exporters—and caused considerable resentment toward the United States. To many North Americans the policy was laudable; to many Latin Americans it was demeaning and paternalistic."[33]

What took place in Latin America during the Nixon years cannot be said in any sense to have been the result of an administration arms transfer policy. To the contrary, much of the effort in this realm was directed to pleading for release from arbitrary ceilings and other blanket restrictions that hampered the development of any meaningful policy toward individual nations in the region on arms transfers. The results were, as we have seen, damaging in political terms to relationships that had been built up over long periods of time and that were advantageous to the United States, replacing cooperation and goodwill with bitterness and resentment. Whatever leverage derives from being the primary arms supplier, not just to a given country but throughout a region —and in the event of hostilities or the threat of them it can be substantial—was forfeited. The advantages from the standpoint of hemisphere defense of standardization of major weapons systems between the United States and Latin American countries were eroded very rapidly. And the first opportunities for Soviet presence in the hemisphere, other than in Cuba, were provided. All in all, it was not a happy result for United States interests.

The provision of arms to South Vietnam during the course of active warfare in which the United States joined it as an ally was, as previously indicated, determined to be outside the scope of this inquiry. Such arms shipments are not security assistance or arms transfers in the usual sense of those terms. Many of the liabilities postulated by critics of arms transfers (most prominently the potential for U.S. involvement in the event of hostilities) simply did not apply, in this case because they had already come about (as a result of a conscious policy decision, it should be noted, not as a by-product of the provision of arms). And of course such warfare was in progress at the time the Nixon administration came into office.

Thus, in trying to get a more meaningful picture of the arms transfers authorized and accomplished by the administration, we have factored out those arms provided to South Vietnam and other primary participants in the war in Vietnam. The intent was to picture the arms transfers that occurred during the period and that were not war-induced, a more meaningful landscape to survey in order to determine the administration's arms transfer policies and their effects.

One of the difficult and long-standing problems the Nixon administration inherited in the realm of arms transfers was the embargo on shipments of arms to South Asia, which the United States had imposed when India and Pakistan went to war over their rival claims to the Kashmir region in 1965. Even in the prior administration it was recognized that such an embargo presented the imposing nation with serious difficulties. For one thing, as Townsend Hoopes told a Senate subcommittee in explaining why the United States lifted the embargo on "nonlethal" items of military equipment the next year, "a policy of total arms suspension was dissipating our influence and producing side effects of serious concern. . . . a policy of total arms suspension made it increasingly difficult for U.S. diplomacy to cope" with unfavorable developments. Of this statement George Thayer later observed that it was "perhaps the closest that any American bureaucrat has come to admitting that to control arms races one must sell arms."[34]

While the loss of influence was a disadvantage, another of far greater potential impact was that the effects of the embargo did not fall equally on India and Pakistan. Whereas India had other sources of aid and arms, Pakistan did not. While India had some capacity for indigenous manufacture of arms and set about developing more, Pakistan had none. The Republican minority on the House Foreign Affairs Committee further suggested that, by continuing to provide economic assistance to India, the United States was in effect subsidizing India's acquisition of Soviet arms.[35] Thus, they held, "while relying heavily on U.S. aid to prevent a famine at home, it is reliably reported that the Government of India is acquiring a sizable quantity of expensive, sophisticated arms, including jet fighters, tanks, and submarines from the Soviet bloc." This included establishing facilities to assemble MiG-21 fighters, construct HF-25 helicopters, and manufacture engines and electronic equipment.[36] The situation had by the autumn of 1968 become so obviously unfair to Pakistan that the Johnson administration, still unwilling to lift the embargo completely, was reported to be actively helping in the search for a third-country source of used tanks for the Pakistanis.[37] India, meanwhile, had with Soviet help built a tank factory said to be capable of producing 360 tanks a year, or about what the United States was able to build at that time.[38]

Within a few months of taking office, the Nixon administration had become concerned about two aspects of this situation: the growing disparity between the military capabilities of India and Pakistan, which was obvious; and the implication of the growing dependence of India upon the Soviet Union for arms. A press report held that "U.S. officials think that in 10 years, perhaps sooner, India's entire defense capacity could be almost totally dependent on Russia for supply, maintenance, and production."[39]

In Pakistan, where it was reported that the Defense Intelligence Agency was in the process of dismantling the electronics complex near Peshawar, which the United States was said to have used in monitoring Soviet missile development, because Pakistan had refused to extend the lease for operation of the base, American officials were resigned to the fact that Pakistan would attempt to diversify its arms suppliers so as to lessen the possibility of being cut off from needed military wherewithal as the result of a supplier-imposed embargo.[40] The extent to which this was pursued, by India as well as Pakistan, became clear when the 1974 Arms Control and Disarmament Agency report indicated that both India and Pakistan had obtained aircraft from six different suppliers.[41]

Even elements of the Congress became concerned about the course of events, with the House Foreign Affairs Committee urging the administration in November 1969 to reconsider the ban on providing lethal equipment to India and Pakistan. In its report on that year's foreign aid authorization, the committee stated that "it may very well be in the national interest to consider selective shipments of military weapons and equipment to both countries."[42] Nearly a year later, in October 1970, the United States announced a decision to sell Pakistan a quantity of arms as a "one-time exception" to the embargo, a point that was made to India at the time. The sale turned out to include some 300 armored personnel carriers, eighteen F-104 Starfighter interceptor aircraft, and seven B-57 Canberra bombers.[43] India predictably "took exception" to the decision, as Secretary Rogers observed in his foreign policy report after the close of the year, even though the sale involved "a limited quantity of military items" and was not to establish a new, continuing supplier relationship.[44] The sale amounted to some $15 million, which compared with an estimated $730 million in arms supplied to India by the Soviet Union since the 1965 war.[45]

In 1971 the United States again imposed a total embargo as dissident elements in East Pakistan engaged in civil war with the central Pakistani government. This blocked the 300 armored personnel carriers which, while part of the one-time sale of the preceding year, had not yet been delivered. These remained in limbo until 1973, when the administration announced that they would be released to Pakistan.[46] When war between India and Pakistan again erupted in December 1971, the United States arranged for Jordan and Iran to transfer to Pakistan some F-5 fighter aircraft to supplement the obsolete Korean War–vintage F-86s that were their principal fighter aircraft.[47] The preceding April, when signs of war began to appear, the United States had shut down the shipment of some $35 million in arms from the prior year's sale, continuing with shipment of $5 million in spare parts that were in the pipeline.[48] The president reported the resumption of fighting between these two South Asian neighbors as one of the "sharp disappointments" of the year.[49] Mr. Nixon pointed out the embargo efforts the United States had employed, the strenuous diplomatic representations by which it had sought to avert fighting—to all of which India had remained indifferent—and the $4.2 billion in economic aid the United Stated has provided to India over the six years since the 1965 war, along with $1.3 billion of the same to Pakistan.[50] Later, someone who had worked in the Defense Security Assistance Agency noted a significant difference between this war and the last: "In the first Indo-Pak war (1965), the US was able to turn it off in 30 days by stopping spare parts shipments to both sides. Now the Soviets and Chinese are filling some of the vacuum and the Europeans the rest. We no longer have that kind of control."[51] The president reflected on the U.S. arms embargo over the years since the 1965 war, observing that it had a much greater effect on Pakistan than on India, so that "the military balance shifted decisively toward India between 1966 and 1971."[52] The result: "It was a foregone conclusion that if war broke out, India would win."[53]

Despite these results, the Senate the following summer enacted into law a provision drafted by Senator Frank Church that banned all military assistance and credit sales of military goods to the South Asian region, including India, Pakistan, and Bangladesh. Only modification on the Senate floor prevented passage of a measure that would also have blocked cash sales.[54] Senator Church, addressing on the Senate floor "this important policy revision, which I authored," told his colleagues of his conviction that "such an embargo on weapons supplied by our Government may help to discourage a new arms race between Pakistan, India, and Bangladesh, where inflated defense budgets and periodic warfare have taken a high and tragic toll."[55]

In March 1973 the president made a decision to release to Pakistan, for what officials described as "more . . . political and psychological reasons than for military ones," something over a million dollars' worth of equipment, which had been ordered in 1970 but blocked by the 1971 reimposition of the total embargo, and return to a policy of permitting spare parts and nonlethal items to be purchased. This decision resulted in the release of the 300 armored personnel carriers that had been for so long in suspended animation. An interesting definitional point was raised in connection with the announcement of the decision, as officials observed that it would be possible to sell ammunition to India or Pakistan, since by Pentagon definitions ammunition could be regarded as "nonlethal" equipment.[56] One's first reaction was to take it as a commentary on recipient-nation marksmanship.

Pakistani Prime Minister Ali Bhutto visited the United States in September 1973, asking that the arms embargo be lifted; after the visit the administration announced that the ban would be continued.[57] But in May 1974 India exploded a nuclear device, and Pakistan's concerns multiplied. One recalled Irving Kristol's question of six years earlier: "Can anyone doubt that—dominoes or no dominoes—the immediate consequence of an American withdrawal from Asia will be India's arming itself with nuclear weapons?"[58] In September India "swallowed up" the kingdom of Sikkim, and Mr. Ali Bhutto took to the public his appeal for the United States to provide Pakistan with arms for its defense: "We stand vindicated in our analysis. Half of our country is gone, half of Kashmir is gone. They marched into Goa and took that. They have gone nuclear. And now they have swallowed up Sikkim."[59] Earlier the Pakistani premier had argued that Pakistan was the only ally to which the United States was refusing to supply arms.[60] Finally, in February 1975, the United States again lifted its embargo on arms sales to India and Pakistan, although no major sales of arms took place throughout the remainder of the administration.[61]

Less than two years after President Ford left office, the Pakistanis had considerable additional reason for concern. Former Prime Minister Bhutto was in prison, accused of the murder of a political rival; long-standing problems with India on the east continued; separatist movements agitated in Baluchistan and the Northwest Frontier Province; a coup d'etat had brought a pro-Soviet regime to power in Afghanistan to the north; and antigovernment rioting that would eventually topple the shah had destabilized the situation in Iran to the west. Little wonder that, to quote an official in Islamabad: "Viewed from here, the world scene looks badly unsettled."[62]

The embargo aspect of arms transfer policy illustrates all the dilemmas that make for so much difficulty and uncertainty in trying to devise and follow a coherent policy in this realm. Not wanting to fuel an arms race, the United States was yet unwilling to have a continuing embargo act to the grave disadvantage of one contending state. Anxious to be on good terms with both, we were frustrated by the ineffectuality of very substantial economic assistance in ameliorating their long-standing antagonisms toward one another. Concerned to arm these states against potential Chinese aggression, we were yet fearful that they would again employ those arms against one another. Desirous of avoiding responsibility for contributing to a conflict situation, we were yet regretful over the loss of influence resulting from the refusal to provide arms. And, determined to block increasing Soviet influence in the region, we were unable to do so without resuming the arms-supplier relationship. All of these seemingly irreconcilable conflicts, compounded by the sometime involvement of the Congress, but not in this case consistently on one side of the issue, made it an agonizing chapter for U.S. policy makers. If nothing else, it demonstrated that, while there are a great many situations in which carefully devised arms transfer strategies can do much that is useful, there is no guarantee that in every case that will be so, and not every problem necessarily has a solution.

Taiwan, Korea and the Philippines, each bound to the United States by a mutual defense treaty, provided important military base facilities, support during the conduct of the war in Vietnam, and residual focuses of interest and influence in the Pacific area under the Nixon Doctrine. Each had been helped importantly by U.S. military assistance, with the Korean armed forces in particular having been wholly rebuilt along American lines following the Korean

War, and with a remarkable degree of success that was demonstrated by the capable divisions the Koreans later sent to assist the allied effort in Vietnam. Taiwan became an extremely important military depot complex, where rebuilding facilities that could perform near miracles on war-damaged equipment operated at full blast during the peak years of the war in Vietnam. The long-standing involvement of the United States in the Philippines was strained by the nature of the Philippine government, particularly when martial law was imposed for an extended period, but the importance of U.S. access and the desire to exercise a moderating influence combined to lend continued importance to that nation in U.S. eyes.

Two weeks before Mr. Nixon took office, a press analysis of events in the Philippines observed that United States–Philippine relations "face a period of strain and abrasive readjustment."[63] Looking back, that appears to have been one of the principal continuities of the Nixon years. Foreign Secretary Carlos Romulo took the occasion of an impending change in the U.S. government to assert that the Philippines wanted to reduce the twenty-five-year lease on U.S. bases, to complain that the Spanish had a better deal in exchange for their base rights, to suggest that the whole matter of the necessity for the bases and indeed of the Philippine military alliance with the United States should be reexamined, and to express the belief that "the United States was no longer dependable as an ally."[64] This was the opening salvo in a running controversy that seemed at bottom to be motivated by the desire to convert base rights into more compensation.

The Republic of the Philippines was the only nation with which the United States had a mutual defense treaty that demanded payment for bases used for that mutual defense. President Marcos demanded a new agreement, to run for five years, with $1 billion compensation in military and economic assistance. The argument continued, sporadically, from this 1972 demand until the very end of the Nixon/Ford period when, in December 1976, Secretary Kissinger offered to make that agreement. But at that point President Marcos turned it down. Presumably he thought he could get even more from the next administration.[65] Military assistance to the Philippines held steady throughout the period at a modest level almost exactly matching the average that had been provided annually since 1950.[66] The bases and presence were maintained, which may be about all that anyone could expect to have been accomplished given the regime with which one had to deal.

The next step after Congress had acted to assume greater control over arms transfers resulting from sales or military assistance was to seek to pass on transfers of U.S. equipment declared to be "excess stocks," which could then be transferred to other nations. As security assistance was progressively reduced, the transfer of excess defense stocks became in certain cases far more important than in the past. Taiwan was one such case. In 1969, for example, the United States agreed to provide Taiwan with a squadron of F-104 jet fighter aircraft declared excess by the U.S. Air Force, thus becoming available for transfer. This enabled Taiwan to replace a squadron of obsolete F-86 aircraft of Korean War vintage.[67] Other equipment, including ships and tanks, which had likewise been declared excess stocks, was also provided to Taiwan.[68] With the establishment of the new relationship between the United States and the People's Republic of China, Taiwan understandably became greatly concerned about its future security and, as has been discussed elsewhere, sought to develop at least the beginnings of an indigenous arms manufacturing capability as a hedge against the loss of external sources of supply, of which the United States was naturally by far the most important. Beginning with the fiscal year following President Nixon's trip to China, Taiwan increased its military sales agreements with the United States nearly threefold, and the average over the next three years was more than double the level of the last previsit year.[69] Military assistance to Taiwan continued for two post-trip years, although the level was only a small fraction of the amount of the sales agreements, then dropped to virtually nothing during the last two years of the administration.[70]

While the level of arms provided was clearly adequate, given sales that supplanted the grants and the transfers of excess defense equipment that had earlier sustained Taiwan, deliveries on the sales agreements naturally lagged years behind the agreements themselves, and concern for the political future of Taiwan led to increasing references to it as one of the so-called pariah states, to frequent discussion of the possibility of nuclear proliferation with respect to Taiwan, and to Taiwan's own efforts to diversify its sources of supply and to build its ability to manufacture armaments. Progressive reduction of the U.S. troop complement stationed on Taiwan served to underscore the changing political realities. Nevertheless, while the Nixon administration continued in office, Taiwan was provided a reliable and adequate supply of armaments for its security. It was not dissatisfaction with performance but apprehension about what the future might hold that led to the search for alternatives.

The Republic of Korea provided a large and active component to the allied forces deployed in Vietnam, so that a good deal of the equipment provided to it during the period of the war was in effect used up. The close relationship between U.S. and Korean units that had existed since the Korean War was reflected in the organization, doctrine, and equipment of Korean forces. More than 50,000 American troops had been stationed in Korea since the war there. Efforts to upgrade the military equipment in the hands of Korean troops received two distinct thrusts; the first stemming from their prospective deployment to Vietnam, where it was thought they could not be expected to fight alongside American troops and have inferior equipment; the second the result of subsequent decisions to reduce the American presence in Korea, so that upgrading of Korean forces was thought prudent to maintain the deterrent and defensive capability in the wake of the U.S. drawdown.

In addition to the provision of increments of first-line equipment such as F-4 fighter aircraft, plans were made to establish a plant in Korea for manufacture of the M-16 rifle, which was being used by U.S. troops in Vietnam.[71] Announc-

ing that the United States had reached a decision to withdraw 20,000 of its troops from Korea by July 1971, Secretary
of State Rogers said that "Korea can properly be regarded
as a good test of the Nixon Doctrine's policy of having local
forces assume additional responsibility for local defense."[72]
At the same time, the United States established a five-year
modernization program for Korean forces, to be funded at
$150 million a year.[73] Eventually the size of that program
was increased to a total of $1.5 billion.[74] A later follow-on
South Korean Force Improvement Plan covering the years
1975–1980 was also devised, and South Korea further determined to become as self-sufficient in arms manufacturing as
possible, although it recognized that complete self-sufficiency in defense production would not be achievable.[75]

During the course of the Nixon administration, the diffusion of arms largely—with the exceptions of Nigeria and
South Africa—bypassed Africa, which was then viewed as
"statistically a military vacuum, possessing very small inventories of military equipment and very low (though increasing) budgets."[76] Certainly this applies to the
arms-supplier role of the United States with respect to
Africa, for during the entire period arms transfers to Africa
amounted to no more than a fraction of 1 percent of total
arms sales and a fraction of 1 percent of military assistance
worldwide as well.[77] Even in the case of South Africa, which
was under international embargo but which the administration wished to influence, according to one account, "by selective relaxation of our stance toward the white regimes, [thus
encouraging] some modification of their current racial and
colonial policies and through more substantial economic assistance to the black states help to draw the two groups
together and exert some influence on both for peaceful
change," the only things sold were dual-purpose (civil and
military application) items, such as helicopters and small
executive airplanes, and herbicides and defoliants.[78]

Only in Ethiopia was there a military assistance program
of any significance, and that held steady at about $10 million
a year until FY1976, when it fell off rapidly. When Secretary

of State Rogers visited Ethiopia in early 1970, a reporter accompanying the party filed an account reflecting Ethiopia's concern about encirclement by Soviet-supported regimes in the Sudan, Somalia, and South Yemen, which led to a request to the United States for more arms. But, he continued, "the attitude of the Nixon administration that [Secretary] Rogers conveyed here today is serious doubt that greater military assistance will solve Africa's problems. The American theme is that security is best assured by internal development, not arms." To this an Ethiopian official responded: "The Russians are moving in and you Americans are moving out."[79] Eventually, of course, the Cubans moved in as well, perhaps giving the secretary some later cause to reconsider whether the development theme by itself had been the wisest course for Ethiopia. John Kennedy had once observed that "diplomacy and defense are not substitutes for one another."[80] This was of course something the Nixon administration knew very well. But neither, as it turned out, were development and defense substitutes for one another, particularly where an outside threat was involved.

What arms transfers there have been to the countries of Africa have not been the subject of much systematic research, a gap noted by two political scientists who have applied statistical analysis techniques to arms transfers to Africa and reached some surprising conclusions. Observing that most of the commentary has in the first instance been concerned with "push factors," or the reasons why suppliers sell or give arms to African states, they took a close look at the "pull factors," the reasons recipient states seek to obtain arms. Their analysis dealt with the influence on arms acquisitions of "the availability of economic resources, defense expenditures, regional arms rivalries, political instability, civil strife, type of regime, and decolonization."[81]

A principal finding was that countries seek arms for reasons that, in Africa at least, are primarily determined by internal aspects of their domestic situation. This suggests that the basic decision that may be influenced by outsiders is not whether a given country shall buy certain arms or a certain level of armament but rather where it will buy what it decides to have. Pointing out that their findings contradict the thrust of much of the arms transfer literature, which is concerned to place responsibility for arms transfers on aggressive sales approaches of supplier nations, the authors conclude that pull factors are three times as important as push factors in determining arms transfers to Africa and that "the most significant influence by far is the level of economic resources available to arms purchasers."[82]

While this study has been concerned with overt arms transfers taking place under military sales and security assistance programs, the episode involving covert provision of some military aid to factions in the Angolan fighting deserves some mention. Both because it became public and led to controversy of the kind that surrounded arms transfers in general, and because the nature of the controversy late in the administration reveals the degree of alienation that had by then developed between executive and legislative elements, Angola provides some interesting insights.

The struggle for political and military control in Angola, which had gone on at lesser levels for many years, reached greater intensity in 1975 when a coalition government was formed for transition purposes, pending independence of Angola from Portugal, which was scheduled for November 1975. By midsummer of that year United Press International reported out of Lisbon that "Soviet ships have docked in Angola with loads of armored cars and heavy weapons for use by the Popular Movement against rival forces in the west African territory," according to refugees who said the materiel was marked as medical supplies.[83] Early in 1975, according to later press reports, the United States had provided funds for political action on the part of the faction it backed, the National Front for the Liberation of Angola (FNLA). One account has it as follows: "Not until July [1975], when Kissinger believed the Vietnam tumult in the United States had subsided enough to risk it, did the Ford administration agree to make its first major investment in arms aid and supplies for the Roberto and Savimbi factions in Angola. It was this secret decision, to channel $14 million in aid to the anti-Communist forces in Angola through Zaire ... that aroused deep dissent inside the Ford administration.... First shipments began arriving in August."[84] Expenditures for arms reached $32 million by November, when an administration request to the Congress, where eight different committees had been briefed on the actions being taken and their progress, caused disagreement with the course of action that eventually became public.

Ultimately the Congress voted on 19 December 1975 to cut off any further funds for support of the Unites States–backed faction in Angola. Commentary in the world press provided an interesting counterpoint to the domestic dialogue in the United States. The *Times* of London, observing that "there is very little point in detente if it does not restrain the Russians from adventures such as they have engaged upon in Angola," said that the Russians and their Cuban allies were "mounting a direct military challenge to the wishes of the majority of Angolans, to the policies of the Organisation of African Unity, and to the interests of the United States."[85] In a commentary that emphasized the significance of arms transfers as a policy instrument, Henry Brandon, the Washington correspondent of London's *Sunday Times,* said that President Ford's reaction to the Senate vote reflected the feelings of Secretary Kissinger, "who believes that the Senate has deprived him of *his most powerful diplomatic tool.*"[86]

In late January 1976 President Ford sent a letter to the Speaker of the House, a letter made public by the White House, expressing his "grave concern over the international consequences of the situation in Angola." After reviewing the course of events in that country over the past year, particularly the intervention of some 10,000 Soviet-backed Cuban combat troops, and pointing out that "the matter of our assistance to Angola was the subject of 25 separate contacts with eight Congressional Committees," Mr. Ford stated his belief that "resistance to Soviet expansion by military means must be a fundamental element of US foreign policy."

He therefore asked the House to disagree to the Senate action barring use of further funds for assistance to factions in Angola.[87] "Brushing aside," as one report said, the president's plea, the House voted 323 to 99 to cut off the aid.[88] In a comment indicative of the state of affairs within the American government at that point, the recipient of the president's letter, Speaker Carl Albert, said that it was "a typical Ford operation—wave your hand, make a gesture and that's the end of it,"[89] further illustrated by Congressman Andrew Young's observation that one need not worry too much about Russian successes in Africa, which would not be lasting, since "the Russians are worse fascists than the Americans."[90]

Geoffrey Kemp later called the Soviet infusion of arms into Angola "the most dramatic effect of arms transfers in recent memory,"[91] a judgment that underscores the importance of the arms differential as a meaningful indicator of impact, far more so than absolute dollar amounts or even numbers of systems. Since the level of armaments in Africa was so low, and since skills in the use of those arms that did exist were in all probability even lower (meaning such elements as training, command and control, leadership experience and ability, organization, combat experience, and the like), the insertion of trained and experienced combat units of Cubans had an effect far greater than would have been the case in a more militarily sophisticated context. Whatever the prospects for countering that influence might have been, the Congress was in no mood to find out, and rejected the president's request for funds by a large majority.

10. Conclusion

The capability of defending itself is at the very heart of a nation's concerns. Its survival, its sovereignty, and the perpetuation of the government in power all rest (in the last resort) on the means to resist external and internal threats. These means are, above all else, weapons. Nations will acquire weapons, no matter what the cost or the impact on their economy or the burden on domestic needs, if they feel themselves threatened. This is what Ali Bhutto meant when he said that Pakistan would eat grass to equal India.

It has been easy for the United States to ignore this fundamental reality. Rich and technically and industrially sophisticated, not since its Revolutionary War has the United States had to look abroad for the military weapons, equipment, and training it felt it needed. The span of two centuries has dulled the perception of how critical that need can be and, in consequence, how strong a threatened nation's determination to acquire arms. Understanding the foregoing is essential to a meaningful analysis of United States arms transfer policy. Aspirations for control of the international traffic in weapons that ignore the powerful impetus of the need for security simply have no contemporary relevance.

Perhaps the first and most obvious factor that comes into play in determining arms transfer policy is that it is not made in isolation. Just as it must be compatible with and serve the interests of the larger foreign policy of the nation, so it must be responsive to events abroad, both contemporary and prospective. This means that what other nations are doing in the way of providing arms and what potential recipient nations are doing about developing indigenous capacities to manufacture arms are important considerations in the formulation of one's own policy. These are but two of the most prominent among a much larger range of relevant factors, most of which are difficult to assess and somewhat indeterminate in their impact, especially in combination.

Thus those who formulate arms transfer policy must in the process deal with a complex range of variables and make necessarily conjectural estimates of likely results. While policy makers sought to deal with these realities, much of the dialogue on arms transfers during the Nixon era had to do with the desirability, feasibility, methodology, and achievement of restraint in the transfer of arms. On each of these aspects there was wide diversity of outlook as to what could and should be done, and even as to what had already been accomplished. The key question was whether restraints could in fact be imposed, for if that were not possible all other related questions become of only theoretical interest.

National attitudes are the crucial determinant of feasibility in attempting systematic restraint of arms transfers. Much depends on the world view of the states involved. Those nations largely satisfied with the current configuration of states and the course of international affairs as they are affected by it naturally favor measures tending to stability and to damping the forces of violent change. But whether that outlook would foster a bias in favor of restraining arms transfers depends on an assessment of their projected effect. Restraint of transfers to revolutionary states, which could be expected to use the arms to disrupt the international order, would obviously be favored. But just the opposite would be the case with arms transfers to states threatened by revolutionary neighbors and therefore in need of arms to defend themselves and their interests. In short, arms have their uses, and attitudes toward restraint of arms transfers vary with a nation's view of the desirability of those uses as practiced by the recipient. The simple significance of the point is that there exists no unanimity of view among states as to whether arms transfers should be restrained.

During the years of the Nixon administration virtually all concern with restraint centered on supplier-imposed limitations, as there was no sustained sentiment for collective restraint on the part of recipients. This is not to say, of course, that recipients did not exercise unilateral restraint; many did. The United States as supplier also routinely imposed restrictions upon arms recipients as to the uses to which military goods could be put, and on further transfers of them, restrictions that have largely been observed. Restraint in contemplating transfers of weapons has gone beyond just the basic decision as to whether to meet a given request (many were not approved) to determination of the number of weapon systems to be provided, their specific configuration and armament, the timing of their delivery, and the composition of related support packages and ammunition supplies.

Beyond restraints of this nature a volume produced for the Council on Foreign Relations as part of its 1980s Project offered some revealing, if pessimistic, insights into the prospects for reducing the international traffic in conventional arms. Forthrightly acknowledging that "it is difficult to get agreement among individuals, let alone governments, as to the nature of the arms trade 'problem' or even whether there is a problem," the report suggested in looking to the coming decade that "the Third World's demand for arms . . . will be too massive and too deeply rooted to be affected by producer restraint."[1] These were, furthermore, clearly not neutral essays, nor were they intended to be: One author was charged with "making a case" for supplier restraint, another with "presenting an advocate's brief" for restraint by arms recipients.[2] Even so, the result is apologetic in tone and overall impact, saying more clearly than any specific argument could that the tide is flowing strongly in the other direction, and for a number of very good reasons that cannot be wished away. So it had been for the preceding decade as well.

Under such circumstances what the issue really amounted to was whether the United States should impose upon itself greater unilateral restraint in refusing outright to provide arms to others. Many argued that the United States should take that approach, although George Thayer's comprehensive survey would shortly lead him to conclude that "it is a fact of life today that no one wants to control the trade in conventional arms."[3] One of the most unequivocal statements on the issue was provided by Paul Warnke during his service in the previous administration as an assistant secretary of defense: "We are no more in a position to cut off sales to our friends abroad than we are to disarm unilaterally. In fact, that would amount to a degree of unilateral disarmament."[4]

Providing a theoretical explanation for the trends, Marshall Shulman had pointed out in a special issue of *Daedalus* devoted to arms control that "the dominant trends in the international system are toward disintegration into violence and anarchy, and any approach to arms control that does not take sufficient account of this central fact is bound to be of limited relevance."[5] But whether at the level of theory or that of practice, the inescapable reality was that where one nation decided unilaterally to withhold arms, others were there to provide them instead. U Thant had said in a United Nations publication of 1970 that "progress toward disarmament, because of its complex nature, can be successfully achieved only if there is a strong political will on all sides to reach agreement."[6] Such agreement did not exist in the late 1960s, nor did it materialize during the years in which the Nixon administration had to formulate United States arms transfer policy.

Very much to the contrary, in fact, systemic changes served to reduce the probability of such agreement, making unilateral restraint in excess of that which the United States was already observing an even more futile approach. Negotiated control of strategic nuclear weapons, which was underway during the entire course of the administration, had the effect of making conventional weapons an increasingly important means of influence in international relations, thereby making suppliers even less willing to give up such an instrument of policy.

Systemic factors deriving from the superpower nuclear standoff affected other powers as well. In commenting on essential means of maintaining the strategic balance and deterrence across the whole range of prospective conflict, many analysts have argued that the near impossibility of actually employing nuclear weapons as long as mutual assured destruction remains a reality has made conventional weapons increasingly significant and that maintaining viable conventional forces is necessary to preserve deterrence at that level.[7] While it is not a new insight to observe that the nuclear umbrella cannot forestall conflict at lesser levels, the potential impact of that limitation has grown with the developing Soviet nuclear capability.

This increasing importance of conventional forces has not been lost on third parties, whose concern for a reliable and adequate (in their terms, not those that may be decided upon unilaterally by a supplier) source of conventional weapons cannot help but be heightened by the perception. If principal allies meanwhile transmit numerous signals of reduced willingness to get involved at first hand to help protect the security of threatened allies, those allies' perceived needs for arms are certain to be further stimulated.

The decline in grant military assistance provided to other countries by the United States also contributed to a changed outlook on the part of arms recipients who, as purchasers, were more likely to want to define their own needs, more particular as to how those needs were satisfied, less dependent on any given supplier, and more interested in shopping around. This has been called by some observers "the single most important factor in accounting for quantitative and qualitative increases in arms transfers."[8] The result was that the efficacy of unilateral restraint was much diminished, a result perhaps not foreseen by a Congress that had urged the shift from grants to sales as strongly as it then wished to influence the international trade through unilateral action.

A concurrent and reinforcing systemic change resulted from the increasing sophistication, complexity, and costliness of major weapons systems. This meant that relatively few nations could manufacture such weapons for themselves, so that obtaining them became a matter of going into the international market. The volume of the arms trade was thereby much increased, but the implications could be misleading; rather than signifying entirely a surge in the number of weapons systems being acquired, it merely showed in part that the more complex systems could less often be domestically produced, so that, instead of building them at home (which would have done nothing to the statistics in arms transfers but would have put the weapons in their hands just as surely), many nations had to shop around to get what they needed. If we look at the system as a whole, the number of major weapons systems remained more nearly constant than the increase in international shipments would seem to indicate, because of this decline in the domestic production capability on the part of lesser states. Nevertheless, the number of potential suppliers, while relatively small for certain items, did not become so limited as to restore the utility of unilateral restraint as a means of limiting arms transfers. It is for this reason that those nations lacking indigenous capacity still find that they can locate alternative sources of supply when one seller or another decides not to fill their requests.

Short of not supplying weapons at all, some of the bases for restraint suggested raise more questions than they answer. One proposition often advanced is that a distinction should be made between offensive and defensive weapons systems and that the systems transferred could then be limited to defensive weaponry. One gains the impression that many who take this position may have limited experience with such things as tanks, air defense missiles, or tactical close-support aircraft. For it is the case in the vast majority of instances that the context of employment of weapons establishes whether their use is offensive or defensive, not the inherent characteristics of the weapons themselves.

A case in point in the Middle East had to do with Jordan and its difficulties in sustaining the moderate regime in power in 1970. Syrian tank columns that invaded Jordan were clearly being used as offensive systems; Jordanian armor that opposed them was just as clearly defensive. Likewise, the tactical air cover that King Hussein threw into the battle was defensive as he used it, whereas the same systems could have been used offensively in other contexts. So it was with the air defense missiles the Soviets provided to Egypt, which were mobile and thus could be varied in their role. Set up along the Suez Canal, they provided a defense against Israeli penetration. Moved forward with attacking Egyptian forces crossing the Canal and heading east, they played an important offensive role in defeating the air assets with which Israel sought to stop the advance.

But even more important than the difficulty of making the distinction between offensive and defensive weapons is the effect on deterrence of aggression if recipient nations were to be restricted to receiving weapons systems adjudged to be useful only in the defense. Inherent in such a policy is the assumption that no nation, once attacked, should do more than repulse the invader or, in other words, defend itself. This presents the attacker with the prospect of very limited liability. The worst he must contemplate is failure, meaning defeat of his attack and restoration of the status quo ante. This, one would imagine, would have minimal deterrent effect in many cases. How much better, from the standpoint of maintaining the peace and discouraging aggression, if the party that initiates such warfare can expect not only to be thwarted in achieving its aims but to risk loss of its own territory, imposed reparations, even occupation and directed political reform. That prospect would seem likely to arouse far more caution in the minds of prospective aggressors than the mere chance that they will not be able to improve their lot. Restriction of arms transfers to defensive systems, even if such a definition could be operationalized, would thus appear likely to reduce conventional deterrence and thereby contribute to a higher incidence of aggressive warfare.

Other judgments advanced with respect to the Nixon years seem to reveal a similar lack of insight into the milieu in which weapons were introduced, thinking in strategic rather than tactical terms. Thus the Council on Foreign Relations study suggested that "in the seventies, the bulk of United States arms transfers has gone to countries whose survival is not obviously vital to American security and foreign policy."[9] This proposition, presented without any supporting evidence or rationale, is radically at variance with the strategic perceptions on which the arms transfer determinations of the period were based. For where had the weapons gone? As we have seen, the largest part was transferred to the Middle East, where Iran, Saudi Arabia, and Israel were the principal recipients. The strategic significance of the region, the criticality of its petroleum resources to Western Europe, Japan, and the United States, and the pivotal role played by these three nations in advancing those interests in the region have been rehearsed so often they have become clichés. Next most significant in terms of both the nature and the quantity of arms transferred were those sent to Western Europe. Again, the strategic significance of the nations of the region and the fundamental importance of it to the United States seem beyond question at this point. The only segment of the arms transfers that would appear to answer to the criticism cited is that having to do with the war in Southeast Asia, where we have at least come to hope that the survival of the nations we sought to assist does not turn out to be vital to American security and foreign policy, given our failure there. But those transfers by no means constituted the bulk of United States arms transfers in the 1970s.

If we combine consideration of the theoretical and policy issues involved in arms transfers and the specifics of implementing the policy in various regions and relationships, it is not difficult to understand how Henry Kissinger might have viewed arms transfers as his single most important instrument of diplomacy, that arms transfers might have come to be the central issue of foreign policy in the closing years of the period, and why consideration of arms transfer policy in context provides so many insights into the administration and its times. Virtually no issue on the central agenda was unrelated to arms transfer considerations, and they were of central concern in dealing with the Middle East, the Soviet Union, Western Europe, and—at least in prospect—China.

The use of arms transfers by the Nixon/Ford administration will continue to be of interest to those concerned with this period in American foreign policy, with arms transfers as an area of specialized interest, and with policy formulation and policy analysis.

In combination with diplomatic initiatives and a coherent set of foreign policy goals, arms transfers served the Nixon administration as an extraordinarily useful instrument of policy, perhaps the single most effective means it employed. The administration did not create the international demand for arms, nor was it responsible for the range of political and economic issues in contention that underlay that demand, but it did use arms transfers wisely and creatively, and with restraint, to serve American interests in world affairs and make important progress in resolving some of those underlying problems that give rise in the first place to nations' need to go armed in the world. To make this case it is not necessary to ignore the inherent limitations on what can be accomplished using arms transfers, the imprecision of such a means, the inefficacy of some arbitrary restraints on such transfers, or the domestic problems that arms transfers sometimes incur. But on the record in its entirety, the policy was undeniably effective and of central importance.

The ultimate goals of the administration's arms transfer policy were to retain as much influence for the United States in world affairs as was possible and to use that influence to help shape new relationships that would be more conducive to peace and security. In this it served to advance a comprehensive foreign policy. Seeking to convince the Congress to support the complementary initiatives that constituted the means of achieving his policy aims, the president had said in his foreign policy report after the remarkable "watershed" year of diplomatic activity that was 1971 that "without an understanding of the philosophical conception upon which specific actions were based, the actions themselves can neither be adequately understood nor fairly judged."[10] In the same way an element of that overall approach, the arms transfer policy, can only be understood and judged in light of an understanding of the larger purposes it was designed to serve and the principles upon which the overall foreign policy depended.

On his final evening as president, Mr. Nixon addressed the American people. Reflecting his continuing sense of the continuity essential to an effective American role in the world, of the distance that was left still to go in achieving the nation's aspirations in performing that role, but also of the progress that had been made, he said once again that "we must complete a structure of peace." The arms transfer policy of his administration had contributed importantly to reaching that goal.

Epilogue

Arms transfers continued to be an important, useful, and controversial instrument of policy in the years following the close of the Nixon/Ford administration. President Carter came to office with an announced determination to cut back severely on American involvement in arms transfers, and only four months into his term asserted that arms transfers would henceforth be an "exceptional policy instrument." This meant, he indicated, that the United States would not be the first to introduce advanced systems into a region, that development of weapons systems designed solely for export would no longer be permitted, that coproduction agreements involving significant weapons would no longer be entered into with foreign nations, and that approvals of weapons sales would be tied to the human rights records of recipient governments.

In addition to these constraints, there was to be a reduction of eight percent in constant dollars in the value of arms transferred in Fiscal Year 1978 as compared to the previous year, and a further eight percent reduction the following year. Negotiations with the Soviet Union to explore possible agreement on supplier restraint in the provision of arms to other nations were also initiated, but collapsed after two fruitless years when the United States delegation returned from a key session in semipublic disarray.

From the beginning there were important exceptions to the new policy. The NATO nations, Japan, Australia, and

New Zealand were exempted from the restrictions, for example. And a widespread perception developed that liberties were being taken in the keeping of accounts to prevent the acknowledged sales from exceeding the proclaimed ceilings; at the same time the actual weapons component of total military sales increased to some two-thirds of the overall value, up from one third and one half in the last two years of the preceding administration.

What happened, in short, was that the Carter administration found in practice that arms transfers were an indispensable instrument of policy and that, if American commitments were to be maintained and American interests served in dealing with foreign nations, no drastic departures from prior practice could be sustained. Thus the practice diverged sharply from the new declaratory policy, an outcome that reflected a necessary accommodation of prevailing realities.

Under the Reagan administration the practices of the later Carter years were more or less continued, but without an accompanying disparity between declaratory policy and actual practice. Arms transfers were never reduced to an "exceptional" instrument of policy under President Carter, and under President Reagan it was not asserted that they should be. Instead the stress was upon deterring aggression by arming allied and other friendly nations, strengthening mutual security relationships, demonstrating United States commitments, enhancing stability, and improving United States defense capability and efficiency. In other words, the stress was placed on the positive aspects of arms transfers.

The eruption of armed conflict over the Falkland Islands between Great Britain and Argentina signalled a field day for pundits the like of which had not been seen since the "high intensity" warfare of the 1973 Arab-Israeli war. Once again the utility of major systems, especially capital ships, was called into question in light of the demonstrated ability of relatively cheap missile systems to bring about their destruction, at least under certain circumstances. While the outcome of these debates seemed likely to be as indeterminate as the earlier arguments over the dynamics of tank-antitank and surface-to-air and air-to-ground missile encounters, the impact on weapons acquisition decisions of many nations seemed likely to be substantial.

The uncertainties arising from these events are among a number of reasons to speculate that arms transfers may, in fact, be reaching a peak, to be followed by a sharp and possibly permanent decline. Among the factors having such bearing on the arms traffic are the proliferation of indigenous arms production capabilities, satisfaction of the initial demand for weapons on the part of newly independent states, the advent of new weapons that could be more destructive yet less expensive (such as missiles capable of destroying ships or aircraft costing many times as much), the growing burden of debt service for states that made large arms purchases on credit, and the difficulty undoubtedly experienced by some recipients in maintaining and operating sophisticated weaponry. In their cumulative impact these developments seem likely to dampen the enthusiasm of many nations for procurement of a next generation of even more complex and expensive weapons systems.

The new nations created with the breakup of former colonial empires after World War II were anxious to acquire the military means of protecting themselves. That large initial demand having essentially been met, however, it was reasonable to expect that orders would taper off. When the time comes to order replacements, moreover, the steeply increasing price of follow-on systems may inhibit some purchasers from trading up. The newest generation of fighter aircraft, for example, is coming in at up to four times the cost of the systems it is designed to replace. Fiscal constraints can be expected to influence many nations that do not feel themselves genuinely threatened to moderate their views about the necessity for maintaining large standing forces, especially as the continuing burden of the money and skilled manpower required to operate and maintain sophisticated weaponry really begins to sink in. There may come a realization of such things as the "little-known fact" pointed out by one expert in naval aviation that the costs of an aircraft over its fifteen-year life cycle vastly exceed the acquisition cost, reaching as much as sixteen times the purchase price in some cases. It is not impossible that under such burdens the recently acquired arsenals of some nations will simply rust, rot, and mildew away.

Purchases of additional or replacement weapons could also be severely inhibited by the burden of debt service incurred as a result of credit purchases. As early as 1973 the debt service requirements of some developing countries had reached the point (not just because of borrowing for arms purchases, of course) that they had become net exporters of capital to the developed world. Some estimates now put cumulative Third World indebtedness as high as $500 billion.

With current oil revenues, the arms acquisition programs of a number of Third World states can be expected to come under predictable stress. It has been possible in the recent past, to a far greater degree than was previously the case, to translate a single factor such as a natural resource (like oil) into military power. But this will, in the case of oil, run its course. One need not indulge in very much speculation to realize that the parameters of the traffic in conventional arms will then change radically. The large amounts of money now available for arms purchases by Third World nations derive principally from oil revenues. These may not last the century in anything like their present flows, and they will in any event become progressively less important as time goes by and new sources of energy are brought on line. It is problematical whether investment of excess earnings from petroleum sales can in the meantime be sufficient to yield later income even approaching present oil profits. For most arms importers dependent upon oil revenues, therefore, the end of the oil is likely to mean the end of the ability to afford anything like the volume or quality of arms now being acquired.

The prospects of the oil drying up—which admittedly is not going to happen tomorrow—should not be overemphasized in thinking about the short- or mid-term aspects of arms transfers. But that eventuality will have an additional impact of presently unimaginable dimensions, stemming from the necessary conversion from the arsenals of petroleum-dependent weapons and vehicles to an entirely new or massively modified suit of weaponry designed to utilize some other kind of fuel. With the oil revenues gone, and the predictably high costs of the conversion to entirely new fleets of post-petroleum weapons and supporting equipment, many nations that are now prominent customers are quite likely to be priced right out of the market for advanced weaponry. In fact, it is quite intriguing to speculate on the incidence and outcome of armed conflict in an era characterized by near bloc obsolescence of such systems as jet aircraft, major naval vessels, and main battle tanks.

These several trends may coalesce to produce a crisis in the whole realm of arms transfers. The shift from grant aid to sales was expedited by a large number of long-term loans, many at concessionary rates. But as we have observed, these have accrued and started to come due, and the ability of many of the recipients to pay could conceivably come into question. Furthermore, those that did default would (presumably) be unable to obtain loans for additional purchases of arms. But a return to grant aid does not seem likely, given the attitudes toward arms transfers that have developed, and the continuing difficulty in obtaining broad support for any foreign aid. Thus the arms transfer business could go into a rather sudden decline (not halt, for the developed nations will still need, want, and be able to pay for weapons to modernize their forces, and the developing nations with their growing indigenous arms industries can be expected to trade at a lower level of sophistication and therefore dollar volume). Meanwhile, the weapons already acquired by some of the less developed nations can be expected to continue to age and deteriorate, which they do at a faster rate than those maintained by technologically more advanced nations with larger pools of trained manpower. The period in which nations, rich and poor, advanced and backward, all possessed large arsenals of sophisticated weapons may, in retrospect, come to be seen as a peculiar and relatively short-lived anomaly.

Nothing is sure when it comes to how nations view their security interests, or the resourcefulness and willingness to sacrifice they are capable of summoning when they consider themselves threatened. But it is predictable that those who perceive a threat will do all in their power to prepare to meet it. Thus while the foregoing factors provide the basis for suggesting that the transfer aspect of conventional arms may well have peaked, it must at the same time be recognized that arms transfers will continue to be an important instrument of policy in the dealings among nations.

Appendix: Arms Transfer Figures and Tables

Figure 1. Foreign Military Sales Worldwide, FY1969-FY1976

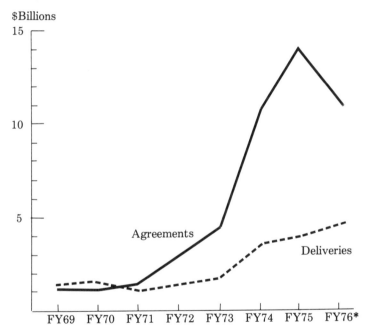

$Billions

Source: Derived from data in U.S. Defense Security Assistance Agency, *Foreign Military Sales and Military Assistance Facts,* December 1978.

Note: Agreements are for the sale of military goods and services, usually in future years and often spread over a period of several years. Not all such agreements are eventually filled, for a variety of reasons. Over the period FY1950–FY1967, for example, foreign military sales

agreements totaled $11,334,596,000, whereas deliveries amounted to
$5,595,199,000, or 49.4% of agreements. Some additional deliveries
under these agreements were probably made in subsequent years, but it
seems likely that deliveries still fell far short of the total agreements, as
for several subsequent years new agreements were running well below
$2 billion annually.

*Adjusted to four-quarter basis using 80% of FY1976/FY197T five-
quarter total. (During 1976 the federal government changed the fiscal
year, which had previously begun on 1 July, to begin on 1 October. To
bridge the period from 30 June 1976, when the last fiscal year beginning
on 1 July ended, to 1 October 1976, when the first fiscal year on the
new schedule would begin, a transition quarter was used, which is
designated FY197T.)

Figure 2. Military Assistance Program, FY1969-FY1976

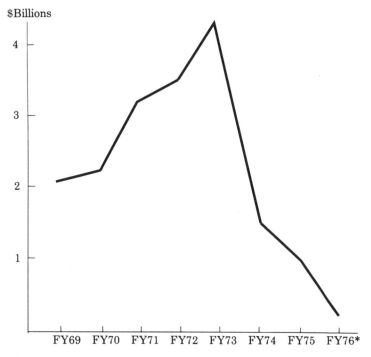

Source: Data in U.S. Defense Security Assistance Agency, *Foreign
Military Sales and Military Assistance Facts,* December 1978.
*Adjusted to four-quarter basis (see note on Figure 1).

Table 1.   Foreign Military Sales by Region, FY1969–
FY1976/197T

	Agreements	
Worldwide	$49,186,748,000	100.0%
Near East & South Asia	32,108,662,000	65.3
Europe	11,434,815,000	23.2
East Asia & Pacific	3,750,519,000	7.6
American republics (less Canada)	804,511,000	1.6
Canada	478,719,000	1.0
International Orgs.	321,796,000	0.7
Africa	287,727,000	0.6
	Deliveries	
Worldwide	$19,557,394,000	100.0%
Near East & South Asia	10,401,871,000	53.2
Europe	6,066,101,000	31.0
East Asia & Pacific	1,863,930,000	9.5
American republics (less Canada)	529,372,000	2.7
Canada	416,980,000	2.1
International Orgs.	220,023,000	1.1
Africa	59,120,000	0.3

Source: Derived from data provided in U.S. Defense Security Assistance Agency, Foreign Military Sales and Military Assistance Facts, December 1978.

Note: Columns may not add due to rounding.

Table 2. Foreign Military Sales by Leading Recipient,
FY1969–FY1976/197T

Rank	Country	$ Total	% of Worldwide	Cumulative %
		Agreements		
1	Saudi Arabia	12,618,995,000	25.7	25.7
2	Iran	11,945,585,000	24.3	50.0
3	Israel	5,383,165,000	10.9	60.9
4	Germany (FRG)	2,718,643,000	5.5	66.4
5	Netherlands	1,437,690,000	2.9	69.3
6	Norway	1,372,821,000	2.8	72.1
7	Belgium	1,347,228,000	2.7	74.8
8	Greece	1,102,847,000	2.2	77.0
9	Australia	1,024,070,000	2.1	79.1
10	Taiwan	986,603,000	2.0	81.1
11	Korea (ROK)	944,843,000	1.9	83.0
12	United Kingdom	686,820,000	1.4	84.4
13	Denmark	678,678,000	1.4	85.8
14	Jordan	632,274,000	1.3	87.1
15	Kuwait	613,441,000	1.2	88.3
16	Spain	570,726,000	1.2	89.5
17	Switzerland	569,997,000	1.2	90.7
18	Canada	478,718,000	1.0	91.7
19	Turkey	444,828,000	0.9	92.6
20	Morocco	431,168,000	0.9	93.5
		Deliveries		
1	Iran	4,346,993,000	22.2	22.2
2	Israel	3,453,568,000	17.7	39.9
3	Germany (FRG)	2,628,803,000	13.4	53.3
4	Saudi Arabia	2,172,592,000	11.1	64.4
5	United Kingdom	1,153,014,000	5.9	70.3
6	Greece	675,979,000	3.5	73.8
7	Australia	600,383,000	3.1	76.9
8	Taiwan	534,181,000	2.7	79.6
9	Canada	416,980,000	2.1	81.7
10	Italy	347,151,000	1.8	83.5
11	Spain	318,918,000	1.6	85.1
12	Korea (ROK)	251,372,000	1.3	86.4
13	Turkey	226,166,000	1.2	87.6
14	Jordan	221,724,000	1.1	88.7
15	Intnl. Orgs.	220,024,000	1.1	89.8
16	Japan	208,393,000	1.1	90.9
17	Norway	203,422,000	1.0	91.9
18	Brazil	177,411,000	0.9	92.8
19	Netherlands	130,134,000	0.7	93.5
20	Switzerland	99,053,000	0.5	94.0

Source: Derived from data in U.S. Defense Security Assistance
Agency, *Foreign Military Sales and Military Assistance Facts*, December 1978.

Table 3. Selected Aspects, Foreign Military Sales Financing
Program, FY1969–FY1976/197T

Total All Methods (payments waived, DOD direct, DOD guaranty)		
Worldwide	$8,684,252,000	100.0%
Near East & South Asia	6,157,331,000	70.9
East Asia & Pacific	985,865,000	11.4
Europe	760,471,000	8.8
American republics	594,442,000	6.8
Africa	136,007,000	1.6
Israel	5,742,664,000	66.1
Greece	450,500,000	5.2
Korea (ROK)	431,958,000	5.0
China (Taiwan)	422,557,000	4.9
Turkey	309,971,000	3.6
Brazil	192,974,000	2.2
Jordan	164,849,000	1.9
Payments Waived		
Worldwide	$2,450,000,000	100.0%
Israel	2,450,000,000	100.0%
DOD Direct		
Worldwide	$2,683,096,000	100.0%
Israel	1,602,664,000	59.7
Greece	158,500,000	5.9
China (Taiwan)	150,157,000	5.6
DOD Guaranty		
Worldwide	$3,551,157,000	100.0%
Israel	1,555,000,000	43.8
Korea (ROK)	370,258,000	10.4
Greece	292,000,000	8.2
China (Taiwan)	272,400,000	7.7
Turkey	190,000,000	5.4

Source: Derived from data in U.S. Defense Security Assistance
Agency, *Foreign Military Sales and Military Assistance Facts*, December 1978.

Table 4. Foreign Military Sales Agreements, Main Battle Tanks

Country	Fiscal Year								Total
	69	70	71	72	73	74	75	76/7T	
Argentina						1			1
Belgium	46								46
Brazil			129	72					201
Chile		40							40
Ethiopia						11	11	14	36
Greece						65	119	23	207
Iran	155			1					156
Israel			250			625	200		1,075
Jordan						130			130
Morocco						26		82	108
Pakistan								30	30
Taiwan					60	100		15	175
Total	201	40	379	73	60	958	330	164	2,205

Source: Compiled from data in U.S. Defense Security Assistance Agency, *Foreign Military Sales: Selected Item Summary*, 10 January 1979.

Table 5. Foreign Military Sales Agreements, Armored Personnel Carriers

Country	Fiscal Year								
	69	70	71	72	73	74	75	76/7T	Total
Argentina		39			53				92
Brazil				500					500
Chile					3				3
Colombia				54					54
Denmark			57						57
Ethiopia							50	27	77
Germany (FRG)				220	300		400		920
Guatemala				5					5
Iran						554	35	83	672
Israel			12	448		3,725	75		4,260
Jordan		100				524			624
Lebanon					50	60			110
Morocco							478		478
Netherlands				42				7	49
Norway						39			39
Pakistan			300						300
Peru						131			131
Philippines							20		20
Saudi Arabia								2	2
Spain	1				151		5	4	161
Taiwan	88	46		114					248
Turkey					132				132
Zaire								2	2
Total	89	185	369	1,383	689	5,033	1,063	125	8,936

Source: Compiled from data provided in U.S. Defense Security Assistance Agency, *Foreign Military Sales: Selected Item Summary*, 10 January 1979.

Table 6. Foreign Military Sales Agreements, TOW Missile

Country	Fiscal Year							Total
	70	71	72	73	74	75	76/7T	
Denmark							100	100
Ethiopia						126	386	512
Germany	3	140	3,500				2,500	6,143
Greece					510	1		511
Iran			8,100		3,800	14,594		26,494
Israel					12,110			12,110
Italy	5				30	4,800		4,835
Korea					664		1,816	2,480
Kuwait				6		2,000		2,006
Lebanon					720			720
Morocco						1,716	460	2,176
Netherlands		17	2,918	5,452		1,400	92	9,879
Norway				16	3,650		2,300	5,966
Oman						200		200
Saudi Arabia							1,774	1,774
Sweden							40	40
Turkey							747	747
Total	8	157	14,518	5,474	21,484	24,837	10,215	76,693

Source: Compiled from data provided in U.S. Defense Security Assistance Agency, *Foreign Military Sales: Selected Item Summary*, 10 January 1979.

Table 7. Foreign Military Sales Agreements, Hawk Missile

Country	Fiscal Year								Total
	69	70	71	72	73	74	75	76/7T	
Greece				4		8			12
Iran					1,788	465			2,253
Israel	120	50	194		36	120			520
Japan					144				144
Jordan								532	532
Spain								144	144
Taiwan							152	190	342
Total	120	50	194	4	1,968	593	152	866	3,947

Source: Compiled from data provided in U.S. Defense Security Assistance Agency, *Foreign Military Sales: Selected Item Summary*, 10 January 1979.

Table 8. Foreign Military Sales Agreements, F-5 Fighter Aircraft

Country	Fiscal Year								Total
	69	70	71	72	73	74	75	76/7T	
Chile						18E			18
Ethiopia								16E	16
Greece							12A		12
Iran	4A	10A		36E	105E	28F			188
Jordan				2B					2
Kenya								10E	12
								2F	—
Korea (ROK)						7B		2A	69
								54E	—
								6F	—
Malaysia				2B					16
				14E					—
Morocco	1B								1
Norway	12A								12
Saudi Arabia				20B			40E		110
				30E			20F		—
Switzerland								66E	72
								6F	—
Taiwan					12B				12
Thailand								13E	16
								3F	—
Total	17	15	—	104	117	53	72	178	556

Source: Compiled from data provided in U.S. Defense Security Assistance Agency, *Foreign Military Sales: Selected Item Summary*, 10 January 1979.

Note: Suffixes indicate various models of the F-5 aircraft.

Notes

PREFACE

1. *Conventional arms* are usually considered to be those which are non-nuclear; chemical and biological weapons are also customarily excluded from the category of conventional weapons. *Arms transfers* is a term used to encompass the various means by which arms are provided to one nation by another; it includes grants, credit sales, and cash sales, as well as certain other specialized but minor categories such as transfer of excess stocks.

1. THE INHERITED SITUATION

1. U.S. Defense Security Assistance Agency (DSAA), *Foreign Military Sales and Military Assistance Facts,* December 1977, pp. 1, 4.

2. Ibid., pp. 17, 19.

3. Ibid., pp. 19–20.

4. Ibid., December 1976, p. 2.

5. *National Observer,* 16 February 1970.

6. DSAA, *Facts,* December 1976, p. 8.

7. Ibid., p. 2.

8. U.S. Congress, Senate, Committee on Foreign Relations, *Arms Sales and Foreign Policy,* p. 3.

9. John Stanley and Maurice Pearton, *The International Trade in Arms,* p. 80.

10. Trevor N. Dupuy and Gay M. Hammerman, eds., *A Documentary History of Arms Control and Disarmament,* p. 121.

11. Ibid., p. 74.

12. United Nations, *The United Nations and Disarmament, 1945–1970,* pp. 29, 37.

13. Ibid., p. 84.

14. Stanley and Pearton, *Trade in Arms,* p. 20.

15. U.S. Congress, House, *Congressional Record,* 10 September 1968, pp. H8447–H8450.

16. *St. Louis Post-Dispatch,* 27 June 1968.

17. *New York Times,* 21 June 1968.

18. U.S. Department of State, Press Release no. 47, 11 March 1968.

19. *Armed Forces Management,* January 1968.

20. *Washington Star,* 17 July 1968.

21. *New York Times,* 18 July 1968.

22. *New York Times,* 30 July 1968.

23. U.S. Congress, House, Committee on Foreign Affairs, *Foreign Assistance Act of 1968,* p. 37.

24. *Washington Post,* 30 July 1968.

25. As quoted in the *New York Times Magazine,* 10 May 1964.

26. In a discussion of international aid, cited in the *Christian Science Monitor,* 18 June 1968.

27. U.S. Military Academy, *Arms Transfers: A Senior Conference, 1976,* p. 59.

28. *New York Times,* 3 April 1972.

29. *Philadelphia Inquirer,* 8 February 1970.

30. *Wall Street Journal,* 19 November 1971.

31. *Business Week,* 7 June 1969.

32. See Ernest Lefever's indictment of CBS television in *TV and National Defense: An Analysis of CBS News, 1972–1973,* and Peter Braestrup's and Burns Roper's two-volume study entitled *Big Story: How the American Press and Television Reported and Interpreted the Crisis of Tet 1968 in Vietnam and Washington.*

33. *Rolling Stone,* 9 February 1978.

34. *Christian Science Monitor,* 2 November 1971.

35. *Mathias Report,* Fall 1978, p. 1.

36. See, for example, Daniel Yankelovich's report, "Cautious Internationalism: A Changing Mood toward U.S. Foreign Policy," in the March/April 1978 issue of the journal *Public Opinion,* and the documentation provided by William Watts and Lloyd A. Free in *State of the Nation, III,* esp. pp. 164–166.

37. John F. Kennedy, in *The Strategy of Peace,* ed. Allan Nevins, as quoted in Alain C. Enthoven and K. Wayne Smith, *How Much Is Enough? Shaping the Defense Program, 1961–1969,* p. 165.

38. Robert Ellsworth, "New Imperatives for the Old Alliance," *International Security,* Spring 1978, p. 135.

39. Data cited by Walter Laquer in "The Psychology of Appeasement," *Commentary,* October 1978, p. 47.

40. *Wall Street Journal,* 30 August 1978.

41. Comparisons derived by economist Warren T. Brooks from data published in the United States Budget for 1979. Reprinted from the *Boston Herald American* in *Human Events,* 9 September 1978.

42. *Washington Post,* 28 May 1978.

43. Henry A. Kissinger, "Reflections on American Diplomacy," as re-

printed in *Fifty Years of Foreign Affairs,* ed. Hamilton Fish Armstrong, p. 257.

44. Ben J. Wattenberg, "Is There a Crisis of Spirit in the West?" *Public Opinion,* May/June 1978.

2. THE NIXON FOREIGN POLICY

1. Lincoln P. Bloomfield and Amelia C. Leiss, *Controlling Small Wars: A Strategy for the 1970's,* p. 393.

2. Linda P. Brady and Nancy K. Barax, "Planning an Arms Transfer Policy for the United States."

3. "Asia after Viet Nam," as reprinted in *Fifty Years of Foreign Affairs,* ed. Hamilton Fish Armstrong, pp. 387–88.

4. U.S. Department of State, *United States Foreign Policy, 1969–1970: A Report of the Secretary of State,* p. 374.

5. Richard Nixon, *U.S. Foreign Policy for the 1970's: A New Strategy for Peace,* pp. 1–2.

6. Ibid., p. 6.

7. Ibid.

8. Ibid., pp. 6–7.

9. Kissinger, "Reflections on American Diplomacy," p. 262.

10. Nixon, *A New Strategy for Peace,* p. 7.

11. Tad Szulc, *The Illusion of Peace: Foreign Policy in the Nixon Years,* p. 124.

12. Ibid., pp. 3, 7.

13. Presidential Message to Congress, "Reorganizing the Foreign Aid Program," 15 September 1970, as cited in Congressional Quarterly, *Nixon: The Second Year of His Presidency,* p. 106-A.

14. *Washington Post,* 3 October 1970.

15. *Philadelphia Inquirer,* 3 May 1970.

16. Richard Nixon, *U.S. Foreign Policy for the 1970's: Building for Peace,* pp. 7, 12, 20, 185.

17. Ibid., p. 166.

18. U.S. Department of Defense, "FY74 DoD Security Assistance Program," *Commanders Digest,* 12 July 1973, p. 6.

19. *Building for Peace,* p. 45.

20. Earl C. Ravenal, "Toward Zero Military Assistance," *Hearings before the Committee on Foreign Relations, United States Senate,* 2–8 May 1973 (reprint of extracted testimony).

21. Congressional Quarterly, *Nixon: The Second Year,* p. 55, citing the president's Budget Message. Spending for human resources was defined as that devoted to education and manpower, health, income security, and veterans' benefits.

22. Ibid., p. 55-A.

23. U.S. Department of State, Press Release no. 337, 8 December 1970, p. 13.

24. *Washington Post,* 6 February 1970.

3. ARMS TRANSFER POLICY AND MECHANISMS

1. The *forward defense countries* now include or have in the past included Greece, India, Iran, South Korea, Laos, Pakistan, the Philippines, Taiwan, Thailand, Turkey, and South Vietnam. The term derives from their geographic locations on the periphery of major communist states.

2. As cited in U.S. Congress, House, Committee on Foreign Affairs, "Statement in Support of the FY 1970 Military Assistance Program and Foreign Military Sales," 24 June 1969.

3. U.S. Department of Defense Directive 5030.28, "Munitions Control Procedures for U.S. Munitions List Export License Applications Referred to the DoD by the Department of State," 10 March 1970.

4. U.S. Congress, Senate, Committee on Foreign Relations, *Security Assistance Authorization: Hearings,* 21 April–2 May 1977, p. 15.

5. International Institute for Strategic Studies (IISS), *Strategic Survey 1976,* p. 24.

6. Craig Powell, "Arms Sales Is More than Just a Military Question," *Armed Forces Management,* January 1968.

7. Presidential Message to Congress, "Reorganization of U.S. Foreign Assistance Programs," 21 April 1971.

8. Lt. Gen. George Seignious, "DOD Organizes to Meet New Challenges of Security Assistance Program," *Defense Management Journal,* April 1972, p. 56.

9. U.S. Arms Control and Disarmament Agency (ACDA), *The International Transfer of Conventional Arms: A Report to the Congress,* p. xi, 29.

10. U.S. Military Academy, *Arms Transfers,* p. 53.

4. POLICY OBJECTIVES

1. For purposes of this discussion, the Middle East is considered to include those nations involved in the broadly encompassing definition used by State Department officials in speaking about the region during this period: Afghanistan, Egypt (the U.A.R), Iran, Iraq, Israel, Jordan, Libya, Oman, Pakistan, Saudi Arabia, South Yemen, Syria, Turkey, the United Arab Emirates, and Yemen were typically included. This largely accords with the configuration of the region designated Near East and South Asia by the

Defense Security Assistance Agency, with the exception of Turkey, which DSAA treats more traditionally as part of Europe.

2. Nixon, *A New Strategy for Peace,* p. 1.

3. Richard Nixon, *RN: The Memoirs of Richard Nixon,* p. 566.

4. Mohamed Heikal, *The Road to Ramadan,* p. 166.

5. Szulc, *The Illusion of Peace,* pp. 90–91.

6. Nixon, *Memoirs,* p. 369.

7. David J. Louscher, "Continuity and Change in American Arms Sales Policies," in Ohio Arms Control Seminar, Workshop II: *Selected Papers,* p. 39.

8. Nixon, *Memoirs,* p. 481.

9. Ibid., p. 477.

10. Ibid., p. 478.

11. Edward Sheehan, *The Arabs, Israelis, and Kissinger,* p. 223.

12. *A New Strategy for Peace,* p. 77.

13. Ibid., p. 81.

14. *Building for Peace,* pp. 123–124.

15. Richard Nixon, *U. S. Foreign Policy for the 1970's: The Emerging Structure of Peace,* p. 140.

16. Szulc, *The Illusion of Peace,* p. 436.

17. Nixon, *Memoirs,* p. 787.

5. EGYPT AS THE KEY

1. Edward C. Luck, "The Arms Trade," *Proceedings of the Academy of Political Science* 32, 4 (1977): 173.

2. ACDA, *The International Transfer of Conventional Arms,* p. 35.

3. Alvin Z. Rubinstein, "Soviet Policy in the Third World in Perspective," *Military Review,* July 1978, p. 5.

4. Ibid.

5. Stanley and Pearton, *Trade in Arms,* p. 207.

6. Heikal, *The Road to Ramadan,* p. 51.

7. Ibid., p. 52.

8. Ibid., p. 64.

9. Ibid., pp. 62–65, 82.

10. Ibid., p. 75. UN Resolution 242 had established the principle that a peace settlement in the Middle East must be based on withdrawal by Israel from territory it occupied in 1967 in return for Arab agreement to end the state of belligerency and to recognize Israel's right to exist and to live in peace. Subsequently UN Resolution 338 established the principle that implementation of Resolution 242, translating the principles it set forth into the specifics of a peace agreement, must be accomplished in negotiations between the parties.

11. Lawrence L. Whetten, *The Arab-Israeli Dispute: Great Power Behaviour,* p. 16.

12. Heikal, *The Road to Ramadan,* pp. 85–90.

13. Szulc, *The Illusion of Peace,* p. 433.

14. Whetten, *Arab-Israeli Dispute,* p. 17.

15. Heikal, *The Road to Ramadan,* p. 17.

16. Ibid., p. 83.

17. Whetten, *Arab-Israeli Dispute,* p. 18.

18. Heikal, *The Road to Ramadan,* p. 96.

19. Ibid., p. 115.

20. Whetten, *Arab-Israeli Dispute,* pp. 18–19.

21. Sheehan, *Arabs, Israelis, and Kissinger,* p. 22.

22. Heikal, *The Road to Ramadan,* p. 120.

23. Ibid., p. 139.

24. Whetten, *Arab-Israeli Dispute,* p. 22, and Heikal, *The Road to Ramadan,* p. 138.

25. Peter Mangold, "The Soviet Record in the Middle East," *Survival,* May/June 1978, pp. 100–101.

26. Congressional Quarterly, *Nixon: The Third Year of His Presidency,* p. 145-A.

27. Ibid., passim.

28. Interview, *As-Sayyad* (Lebanon), 30 December 1976, as reprinted in *Survival,* March/April 1977, p. 81.

29. Jim Hoagland, review of *In Search of Identity: An Autobiography* by Anwar el-Sadat, *Washington Post,* 7 May 1978.

30. Department of State, *United States Foreign Policy, 1969–1970,* p. 80.

31. Heikal, *The Road to Ramadan,* p. 112.

32. Department of State, *United States Foreign Policy, 1969–1970,* p. 80.

33. Ibid., p. 77.

34. Heikal, *The Road to Ramadan,* p. 117.

35. Ibid., p. 118.

36. Ibid., p. 119.

37. Ibid., p. 132.

38. *The Emerging Structure of Peace,* p. 138.

39. Ibid., p. 139.

40. Ibid., p. 134.

41. Robert A. Gessert and William W. Cover, *Qualitative Constraints on Conventional Armaments,* 2: 27.

42. Heikal, *The Road to Ramadan,* p. 167.

43. Whetten, *Arab-Israeli Dispute,* p. 23.

44. Ibid.

45. Heikal, *The Road to Ramadan,* p. 170.

46. Szulc, *The Illusion of Peace,* p. 602.

47. Sheehan, *Arabs, Israelis, and Kissinger,* p. 23.

48. Ibid., p. 26.

49. Whetten, *Arab-Israeli Dispute,* p. 23.

50. Marvin Kalb and Bernard Kalb, *Kissinger,* p. 451. Although the Soviets recovered their sense of proportion enough to come through with arms, thereby protecting, for the time being, their naval base rights and the Treaty of Friendship and Cooperation, the expulsion was a humiliation that has not been forgotten. Even today the memory of their treatment rankles, as an extract from a current lesson plan for indoctrination of Soviet military personnel makes clear: "It is not by chance that the nations of many countries around the world decisively condemned the capitulatory policy of the present leaders of Egypt, who unilaterally abrogated the treaty of friendship and cooperation with the Soviet Union and set a course for open collusion with imperialism in defiance of the national interests of their own people and the interests of the other Arab countries and the people of Palestine." Lt. Col. N. Khibrikov, "Downfall of the Colonial System of Imperialism: Liberated Countries of Asia, Africa, and Latin America," in [U.S.] Air Force Intelligence Service, *Soviet Press: Selected Translations,* August 1978, p. 240.

51. Heikal, *The Road to Ramadan,* p. 181.

52. Ibid., p. 173.

53. Richard Nixon, *U.S. Foreign Policy for the 1970's: Shaping a Durable Peace,* p. 139.

54. Nixon, *Memoirs,* pp. 921–922.

55. Heikal, *The Road to Ramadan,* p. 168.

56. Anthony Nutting, *The Arabs: A Narrative History from Mohammed to the Present,* p. 397.

57. Sheehan, *Arabs, Israelis, and Kissinger,* p. 58 (emphasis supplied).

58. Ibid., p. 66.

59. *Washington Post,* 15 October 1978.

60. *Shaping a Durable Peace,* p. 140.

61. Heikal, *The Road to Ramadan,* p. 270.

62. Ibid., p. 273.

63. Ibid., p. 275.

64. Szulc, *The Illusion of Peace,* p. 762.

65. Heikal, *The Road to Ramadan,* pp. 20–26.

66. Ibid., p. 214.

67. Nixon, *Memoirs,* p. 986.

68. *New York Times,* 17 May 1978.

69. *U.S. News & World Report,* 29 May 1978.

70. Whetten, *Arab-Israeli Dispute,* p. 41.

71. Nixon, *Memoirs,* p. 1011.

72. Whetten, *Arab-Israeli Dispute,* pp. 36–37.

73. *As-Sayyad* interview, in *Survival,* March/April 1977, p. 81.

74. Ibid.

75. These views were revealed in an interview with the Egyptian weekly

October, published in the issue of 25 February 1978, as quoted in the *Washington Post,* 26 February 1978.

76. Anne H. Cahn et al., *Controlling Future Arms Trade,* p. 78.

77. Ibid., p. 85.

78. Sheehan, *Arabs, Israelis, and Kissinger,* p. 92.

79. Cahn, *Controlling Future Arms Trade,* pp. 35–36.

80. Statement before the Subcommittee on Foreign Operations of the Senate Appropriations Committee, U.S. Department of State, Press Release no. 321, 24 July 1974, pp. 1–2.

81. Sheehan, *Arabs, Israelis, and Kissinger,* p. 177.

82. NBC Television Network, 15 August 1974.

83. Sheehan, *Arabs, Israelis, and Kissinger,* p. 76 (emphasis in original).

84. Whetten, *Arab-Israeli Dispute,* p. 34.

85. *As-Sayyad* interview, in *Survival,* March/April 1977, p. 81.

86. *Washington Post,* 25 February 1978.

87. *As-Sayyad* interview, in *Survival,* March/April 1977, p. 80.

88. Heikal, *The Road to Ramadan,* p. 165.

89. Congressional Research Service, *Implications of President Carter's Conventional Arms Transfer Policy,* p. 56.

90. *Washington Post,* 9 February 1976.

91. *New York Times,* 8 February 1976.

92. *Boston Glove,* 1 February 1976.

93. *Congressional Quarterly,* 7 February 1976, p. 298.

94. *New York Times,* 28 February 1978.

6. TRYING TO LIVE WITH ISRAEL

1. *Baltimore Sun,* 17 September 1968.

2. As reported in *Radio-TV Defense Dialog,* 16 September 1968, quoting CBS Television "Morning News."

3. *New York Times,* 15 September 1968.

4. *New York Times,* 21 September 1968.

5. *Department of State Bulletin,* 28 October 1968, p. 452.

6. Ibid.

7. *Baltimore Sun,* 15 December 1968.

8. *New York Times,* 30 December 1968; *Baltimore Sun,* 28 December 1968.

9. IISS, *Strategic Survey 1969,* p. 106.

10. *Washington Post,* 11 January 1969.

11. Golda Meir, *My Life,* p. 385.

12. Ibid., pp. 387–388, 391.

13. Kalb and Kalb, *Kissinger,* p. 191.

14. Heikal, *The Road to Ramadan,* p. 43.
15. Szulc, *The Illusion of Peace,* p. 92.
16. Ibid., p. 93.
17. Ibid., p. 97.
18. Department of State, *United States Foreign Policy, 1969–1970,* p. 74.
19. Congressional Quarterly, *Nixon: The Second Year,* p. 144-A.
20. Ibid., p. 15.
21. *Baltimore Sun,* 13 June 1970.
22. Szulc, *The Illusion of Peace,* p. 333.
23. Ibid., p. 320.
24. Nixon, *Memoirs,* p. 480.
25. Ibid., p. 484.
26. Congressional Quarterly, *Nixon: The Second Year,* p. 15.
27. Szulc, *The Illusion of Peace,* p. 210.
28. U.S. Department of State, Press Release no. 337, 8 December 1970, pp. 3, 14.
29. Whetten, *Arab-Israeli Dispute,* pp. 21–22.
30. Edward N. Krapels, *Oil and Security: Problems and Prospects of Importing Countries,* p. 24.
31. Szulc, *The Illusion of Peace,* p. 441.
32. *New York Times,* 21 January 1972.
33. *The Emerging Structure of Peace,* pp. 136–138.
34. *New York News,* 19 June 1973.
35. Heikal, *The Road to Ramadan,* p. 203.
36. Nixon, *Memoirs,* p. 770.
37. Meir, *My Life,* pp. 415–416.
38. Ibid., p. 417.
39. Nixon, *A New Strategy for Peace,* p. 4.
40. Meir, *My Life,* p. 430.
41. William B. Quandt, "Soviet Policy in the October Middle East War —II," *International Affairs,* October 1977, p. 591, n. 11.
42. Szulc, *The Illusion of Peace,* p. 732.
43. Ibid., p. 735.
44. Nixon, *Memoirs,* p. 922.
45. Heikal, *The Road to Ramadan,* pp. 244–245.
46. Nixon, *Memoirs,* pp. 926–927.
47. Ibid., pp. 927–928.
48. Kalb and Kalb, *Kissinger,* p. 471.
49. Meir, *My Life,* p. 431.
50. Kalb and Kalb, *Kissinger,* p. 501.
51. Nixon, *Memoirs,* p. 921.
52. *New York Times,* 26 January 1975.
53. As quoted in Heikal, *The Road to Ramadan,* pp. 238–239.
54. *Washington Post,* 7 May 1978.
55. Meir, *My Life,* p. 433.

56. Ibid., p. 430.

57. Szulc, *The Illusion of Peace*, p. 738.

58. Nixon, *Memoirs*, p. 984.

59. Krapels, *Oil and Security*, p. 19.

60. Ibid., p. 27.

61. Nixon, *Memoirs*, pp. 1007–1008.

62. Ibid., p. 1018.

63. Meir, *My Life*, p. 442.

64. *Defense Space Business Daily*, 2 January 1976.

65. *New York News*, 12 September 1974.

66. *Washington Post*, 12 September 1974.

67. Sheehan, *Arabs, Israelis, and Kissinger*, p. 69n.

68. Ibid., p. 200.

69. Anthony H. Cordesman, "How Much Is Too Much?" *Armed Forces Journal International*, October 1977.

70. *Washington Post*, 6 December 1978.

71. Meir, *My Life*, p. 450.

72. Whetten, *Arab-Israeli Dispute*, p. 38.

73. Ibid., p. 41. The principal points of the agreement were renunciation of the use of force and agreement to resolution of conflict by peaceful means, provision for passage through the Suez Canal of nonmilitary cargo destined for Israel, a demilitarized buffer zone manned by the United Nations Emergency Force, and a team of American technicians to man an early warning station and electronic sensor fields. In addition, a joint commission was created to deal with problems that might arise in the course of implementing the agreements.

74. Sheehan, *Arabs, Israelis, and Kissinger*, pp. 253, 255.

75. Ibid., p. 256.

76. Joseph J. Sisco, ed., *Prospects for Peace in the Middle East*, p. 15.

77. Sheehan, *Arabs, Israelis, and Kissinger*, p. 163.

78. Ibid., p. 175.

79. *Washington Post*, 31 July 1975.

80. Cordesman, "How Much Is Too Much?"

7. JORDAN AND THE PERSIAN GULF MOSAIC

1. U.S. Department of State, Press Release no. 312, 24 July 1974, p. 2.

2. Heikal, *The Road to Ramadan*, p. 62.

3. Ibid., p. 67.

4. *Los Angeles Times*, 8 April 1971.

5. Szulc, *The Illusion of Peace*, p. 762.

6. Lt. Col. William M. Constantine, "The 'Hawk' Controversy: The Proposed Sale of Air Defense Systems to Jordan," pp. 4–5.

7. Ibid., p. 33.

8. Luck, "The Arms Trade," p. 171.

9. *Radio-TV Defense Dialog,* Broadcasts of 18–19 October 1968, p. 9.

10. *U.S. News & World Report,* 2 October 1978, p. 73.

11. Bill T. Gaugash, "Iran: Strategy for Survival," *International Relations,* Spring 1978, p. 52. Everyone has his issue. Even Senator Fulbright, who had spent a good deal of his time trying to block administration efforts to get involved in one way or another overseas, was moved by the prospect of an oil embargo to issue a colorfully phrased warning: "The Persian Gulf countries," he said, "would be well-advised not to press too hard, and to treat their oil wealth as a kind of global trust, if for no other reason than for their own protection. The meat of the gazelle may be succulent indeed, but the wise gazelle does not boast of it to lions" (quoted in *Army,* February 1974). This he had said in the spring of 1973; when the embargo came, six months later, it was unfortunately revealed as no more than simple bluster. If any further evidence were needed as to how vastly more advantageous it was to have a surrogate such as Iran maintaining the security and stability of the oil-producing region, rather than having to try to figure out how to do something about it ourselves if something should happen to disrupt those necessary conditions, surely the embargo and its aftermath served the purpose. For that was a situation in which the United States still retained some diplomatic leverage, yet which still proved difficult enough. Trying to deal with a similar situation precipitated by a militarily hostile power offered far less happy prospects.

12. Robert G. Irani, "US Strategic Interests in Iran and Saudi Arabia," *Parameters* 7, no. 4, (1977): 21–31.

13. *Baltimore Sun,* 2 June 1969.

14. Szulc, *The Illusion of Peace,* p. 167.

15. Nixon, *Memoirs,* p. 679.

16. Department of State, *United States Foreign Policy, 1969–1970,* p. 85.

17. Ibid., p. 89.

18. The communiqué following the president's visit cited the "vital importance" of the security and stability of the Persian Gulf and spoke of Iran's "determination to bear its share of this responsibility." Szulc, *The Illusion of Peace,* p. 585.

19. *Washington Post,* 7 November 1978.

20. Edward M. Kennedy, "The Persian Gulf: Arms Race or Arms Control?" *Foreign Affairs,* October 1975, p. 21.

21. *Baltimore News-American,* 28 July 1975.

22. *Wall Street Journal,* 23 July 1975.

23. *Department of State Bulletin,* 14 July 1975, p. 81.

24. *Baltimore Sun,* 30 November 1978.

25. *Baltimore Sun,* 30 October 1978.

26. *Washington Post,* 6 November 1978.

27. *New York Times,* 8 November 1978.

28. *New York Times,* 26 November 1978.

29. Imperial Embassy of Iran, "Economic Development of Iran, 1945–1978," *Profile of Iran,* June/July 1978, pp. 7–13.

30. Ibid., pp. 37–43.

31. George Lenczowski, ed., *Iran under the Pahlavis,* p. iii.

32. Ibid., p. xxii.

33. Ibid., p. xv.

34. *U.S. News & World Report,* 13 November 1978, p. 36.

35. Ibid.

36. U.S. Department of State, "U.S. Strategy for Iran (Reply to NSAM 228)," sanitized version, Declassified Document 1976:43A, *Declassified Documents Quarterly Catalogue,* Tab A, p. 1.

37. Ibid., p. 5.

38. Ibid., p. 1.

39. *Washington Post,* 11 November 1978.

40. *New York Times,* 9 July 1978.

41. *New York Times,* 8 November 1978.

42. *Baltimore Evening Sun,* 7 November 1978.

43. *Newsweek,* 11 December 1978, p. 56.

44. *Los Angeles Times,* 5 February 1979.

45. *Time,* 5 February 1979, p. 110.

46. *New York Times Magazine,* 17 December 1978, p. 27.

47. *New York Times,* 22 January 1979.

48. *New York Times,* 11 February 1979.

49. *Washington Post,* 17 January 1979.

50. *New York Times,* 16 February 1979.

51. *Time,* 1 October 1979, p. 33.

52. *New York Times,* 19 February 1979.

53. *New York Times,* 12 November 1979.

54. News conference of 27 February 1979, as quoted in U.S. Department of the Air Force, *Selected Statements #79–3,* 1 May 1979, p. 43.

55. *Richmond Times-Dispatch,* 15 November 1978.

56. *Washington Post,* 13 February 1979.

57. *Washington Post,* 11 December 1979.

58. DSAA, *Facts,* December 1978.

59. Cahn, *Controlling Future Arms Trade,* p. 89.

60. Krapels, *Oil and Security,* p. 21.

61. Department of State, *United States Foreign Policy, 1969–1970,* p. 84.

62. *New York Times,* 5 June 1973.

63. Sheehan, *Arabs, Israelis, and Kissinger,* p. 69.

64. *New York Times,* 11 September 1974.

65. *Department of State Bulletin,* 14 July 1975, p. 75.

66. Ibid., p. 78.

67. See Jim Hoagland and J. P. Smith, "Saudi Arabia and the United States: Security and Interdependence," *Survival,* March/April 1978, pp.

80–81; Krapels, *Oil and Security,* p. 20; and U.S. Congress, House, Committee on International Relations, *Review of Recent Developments in the Middle East: Hearing,* June 8, 1977, p. 117.

68. *Aerospace Daily,* 31 July 1974.
69. *New York Times,* 5 June 1973.
70. Ibid.
71. *Department of State Bulletin,* 14 July 1975, p. 77.
72. Nixon, *Memoirs,* p. 1008.
73. Meir, *My Life,* p. 423.
74. *Newsweek,* 11 December 1978, p. 56.
75. Kalb and Kalb, *Kissinger,* p. 544.

8. NATO AND WEST EUROPEAN ARMS TRANSFERS

1. Cahn, *Controlling Future Arms Trade,* p. 68.
2. Beau Morris, "Why France's Arms Exports Make It a Paper Tiger," *Armed Forces Journal International,* October 1978, p. 19.
3. Ibid.
4. Details of Aerospatiale operations are from the *New York Times,* 28 May 1978.
5. ACDA, *The International Transfer of Conventional Arms,* p. 32.
6. "New Opportunities for the French Aeronautical Industry with Respect to the United States," as summarized by Dr. Joseph Annunziata in the *Friday Review of Defense Literature,* 13 August 1971, pp. 4–5.
7. Michel Theoval, in *Armed Forces Journal International,* October 1978, p. 23.
8. ACDA, *The International Transfer of Conventional Arms,* p. 33.
9. Senate, Committee on Foreign Relations, *Security Assistance Authorization: Hearings,* p. 129.
10. *Washington Post,* 8 November 1971. See also Nixon, *Shaping a Durable Peace,* p. 83; and "Keeping Defense Tuned in with International Issues," *Armed Forces Management,* October 1968, p. 117.
11. ACDA, *The International Transfer of Conventional Arms,* p. 34.
12. Herbert Y. Schandler, "U.S. Arms Export Policies: A Candid Assessment."
13. *Aviation Week & Space Technology,* 2 June 1975, p. 197.
14. *Wall Street Journal,* 6 February 1976.
15. *New York Times,* 5 July 1975.
16. *Washington Post,* 25 November 1978.
17. *Wall Street Journal,* 6 February 1976.
18. *New York Times,* 28 July 1975.
19. Atlantic Council of the United States, Working Group on Security, *The Growing Dimensions of Security,* p. 104.

20. U.S. Department of State, Press Release no. 128, 2 May 1973, p. 6.
21. *Washington Post,* 23 March 1972.
22. *New York Times,* 17 February 1969.
23. IISS, *Strategic Survey 1974,* p. 77.
24. *New York Times,* 12 February 1976.
25. "Turkey's Security Policies," *Survival,* September/October 1978, pp. 203, 205.
26. *Washington Post,* 3 August 1975.
27. Ibid.
28. Cahn, *Controlling Future Arms Trade,* p. 120.
29. *Newsweek,* 11 December 1978, p. 56.

9. LATIN AMERICA, ASIA, AND AFRICA

1. U.S. Department of State, "U.S. Policy re Latin American Military Purchases," Cable #CA-6370 dated 23 December 1963, Declassified Document 1976:97D, *Declassified Documents Quarterly Catalogue,* pp. 1–2.
2. *Baltimore Sun,* 18 May 1968.
3. Harold Brown, "Mutual Security in a Changing World," p. 6.
4. Stanley and Pearton, *Trade in Arms,* pp. 219–220.
5. *Baltimore Sun,* 10 May 1968.
6. Annette Baker Fox, *The Politics of Attraction,* p. 97.
7. Stanley and Pearton, *Trade in Arms,* p. 220.
8. Cited by John L. Hess, "Poor Nations Spend Fortune on Arms Purchases," *New York Times,* 18 August 1969.
9. *Washington Star,* 20 April 1969.
10. *Los Angeles Times,* 11 June 1970.
11. *A New Strategy for Peace,* p. 101.
12. *San Diego Union,* 19 March 1972.
13. U.S. Congress, Senate, *Congressional Record,* 24 July 1972, p. S11663.
14. Ibid., p. S11665.
15. Ibid., p. S11666.
16. *Aviation Week & Space Technology,* 20 March 1972.
17. *Memphis Press Scimitar,* 16 June 1973.
18. U.S. Central Intelligence Agency, *Changing Patterns in Soviet-LDC Trade, 1976–77: A Research Paper,* p. 5 (unclassified publication).
19. Department of Defense, "FY74 DoD Security Assistance Program," p. 4.
20. U.S. Department of State, Press Release no. 128, 2 May 1973, p. 5.
21. U.S. Department of State, Press Release no. 192, 5 June 1973, p. 4.

22. "The Proliferation of Conventional Arms," in International Institute for Strategic Studies, *The Diffusion of Power: I. Proliferation of Force,* p. 40.

23. Louscher, "Continuity and Change in American Arms Sales Policies," in Ohio Arms Control Seminar, *Workshop II: Selected Papers,* p. 40.

24. Congressional Research Service, *Implications of Carter's Arms Transfer Policy,* p. 41.

25. Ibid., pp. 38–39.

26. *Defense & Foreign Affairs Digest,* August 1978, p. 12.

27. Ibid., p. 11.

28. *Defense & Foreign Affairs Digest,* September 1978, pp. 25–26.

29. Gregory F. Treverton, *Latin America in World Politics: The Next Decade,* p. 9.

30. *Trade in Arms,* p. 220.

31. U.S. Department of State, "US Arms Transfer Policy in Latin America," July 1978, p. 1.

32. Congressional Research Service, *Implications of Carter's Arms Transfer Policy,* p. 70.

33. Cahn, *Controlling Future Arms Trade,* p. 43.

34. George Thayer, *The War Business,* pp. 234–235.

35. U.S. Congress, House of Representatives Report no. 1587, 26 June 1968, p. 56.

36. Ibid., p. 55.

37. *Washington Post,* 18 October 1968.

38. *Washington News,* 10 June 1969.

39. Ibid.

40. *Washington Post,* 10 March 1969.

41. ACDA, *The International Transfer of Conventional Arms,* p. 13n.

42. *New York Times,* 9 November 1969.

43. *Chicago Tribune,* 9 October 1970.

44. Department of State, *United States Foreign Policy 1969–1970,* p. 92.

45. Szulc, *The Illusion of Peace,* p. 441.

46. ACDA, *The International Transfer of Conventional Arms,* p. 86.

47. According to the Szulc account, *The Illusion of Peace,* p. 443.

48. Nixon, *The Emerging Structure of Peace,* p. 142.

49. Ibid., p. 11.

50. Ibid., pp. 145–147.

51. *Defense & Foreign Affairs Digest,* September 1978, pp. 16–17.

52. *The Emerging Structure of Peace,* p. 147.

53. Ibid., p. 142.

54. *Washington Post,* 14 June 1972.

55. U.S. Congress, Senate, *Congressional Record,* 24 July 1972, p. S11671.

56. *New York Times,* 15 March 1973; see also Nixon, *Shaping a Durable Peace,* p. 146.

57. *New York Times,* 9 July 1974.

58. *New York Times,* 12 May 1968.

59. *Washington Post,* 11 September 1974.

60. *New York Times,* 9 July 1974.

61. Congressional Research Service, *Implications of Carter's Arms Transfer Policy,* p. 47.

62. *New York Times,* 24 December 1978.

63. *New York Times,* 6 January 1969.

64. Ibid.

65. *Washington Post,* 27 August 1978.

66. DSAA, *Facts,* December 1977, p. 17.

67. *New York Times,* 10 January 1970.

68. *New York Times,* 29 March 1970.

69. DSAA, *Facts,* December 1977, p. 1.

70. Ibid., p. 17.

71. *New York News,* 13 June 1969.

72. U.S. Department of State, Press Release no. 337, 8 December 1970, p. 9.

73. Ibid., p. 10.

74. *Washington Post,* 12 May 1972.

75. U.S. Department of State, *Special Report No. 40,* January 1978, pp. 3–4.

76. Geoffrey Kemp, "Diffusion of Power," in U.S. Military Academy, *Arms Transfers,* p. 80.

77. DSAA, *Facts,* December 1977, pp. 1–2, 17–18.

78. Szulc, *The Illusion of Peace,* pp. 221–223.

79. *Washington Post,* 13 February 1970.

80. University of Washington Address, 1961, as quoted in Theodore Sorensen, *Kennedy* (New York: Harper & Row, 1969), p. 511.

81. William P. Avery and Louis A. Picard, "The Political Economy of Conventional Arms Transfers to Africa," p. 2.

82. Ibid., p. 16.

83. *Washington Star,* 22 July 1975.

84. *Washington Post,* 6 January 1976.

85. U.S. Information Agency, "Senate Votes to Deny Angola Aid," *Media Reaction Report No. 102,* 22 December 1975, pp. 2–3.

86. Ibid., p. 4.

87. White House, Press Release, 27 January 1976.

88. *New York Times,* 28 January 1976.

89. Ibid.

90. Ibid.

91. U.S. Military Academy, *Arms Transfers,* p. 33.

10. CONCLUSION

1. Cahn, *Controlling Future Arms Trade,* pp. 1, 18.

2. Ibid., p. 5.

3. Thayer, *The War Business,* p. 364.

4. *New York Times,* 21 June 1968.

5. Marshall D. Shulman, "Arms Control in an International Context," *Daedalus,* Summer 1975, p. 54.

6. United Nations, *The United Nations and Disarmament, 1945–1970,* p. vi.

7. See, for example, Dennis Ross, "Rethinking Soviet Strategic Policy: Inputs and Implications," *Journal of Strategic Studies,* May 1978, pp. 24–25.

8. U.S. Military Academy, *Arms Transfers,* p. 67.

9. Cahn, *Controlling Future Arms Trade,* p. 34.

10. *The Emerging Structure of Peace,* p. 2.

Selected Bibliography

Armstrong, Hamilton Fish, ed. *Fifty Years of Foreign Affairs.* New York: Praeger, 1972.

Atlantic Council of the United States, Working Group on Security. *The Growing Dimensions of Security.* Washington, D.C., 1977.

Avery, William P. "Domestic Influences on Latin American Importation of U.S. Armaments." *International Studies Quarterly,* March 1978, pp. 121–142.

————, and Louis A. Picard. "The Political Economy of Conventional Arms Transfers to Africa." Paper presented at the 1978 Annual Convention of the International Studies Association, Washington, D.C., February 1978.

Bloomfield, Lincoln Palmer, and Amelia C. Leiss. *Controlling Small Wars: A Strategy for the 1970's.* New York: Knopf, 1969.

Brady, Linda P., and Nancy K. Barax. "Planning an Arms Transfer Policy for the United States." Paper presented at the Annual Meeting of the Southern Political Science Association, New Orleans, November 1977.

Braestrup, Peter, and Burns Roper. *Big Story: How the American Press and Televison Reported and Interpreted the Crisis of Tet 1968 in Vietnam and Washington.* 2 vols. Boulder: Westview, 1977.

Brown, Harold. "Mutual Security in a Changing World." Remarks at the Annual Convention of the Exchange Club, Kansas City, Mo., 29 July 1968, as published in Supplement no. 9-1968 to the *Air Force Policy Letter for Commanders.* Washington, D.C.: Office of the Secretary of the Air Force, 1968.

Buchan, Alastair. *The End of the Postwar Era: A New Balance of World Power.* New York: Dutton, 1974.

Cahn, Anne H., et al. *Controlling Future Arms Trade.* New York: McGraw-Hill, 1977.

Congressional Quarterly. *Nixon: The Second Year of His Presidency.* Also volumes for third and fourth years. Washington, D.C., 1971 and subsequent years.

Congressional Research Service. *Implications of President Carter's Conventional Arms Transfer Policy.* Washington, D.C.: Library of Congress, 1977.

Constantine, Lt. Col. William M. "The 'Hawk' Controversy: The Proposed Sale of Air Defense Systems to Jordan." Washington, D.C.: National War College, April 1976.

Cordesman, Anthony H. "How Much Is Too Much?" *Armed Forces Journal International,* October 1977.

Declassified Documents Quarterly Catalogue. Vols. 1 (nos. 1–4) through 4 (nos. 1–2), January 1975–June 1978. Washington, D.C.: Carrollton Press, 1975 and subsequent years.

Dupuy, Trevor N., and Gay M. Hammerman, eds. *A Documentary History of Arms Control and Disarmament.* New York: Bowker, 1973.

Ecevit, Bulent. "Turkey's Security Policies." Text of an address given to the International Institute for Strategic Studies, 15 May 1978. *Survival,* September/October 1978, pp. 203–208.

Ellsworth, Robert. "New Perspectives for the Old Alliance." *International Security,* Spring 1978.

Enthoven, Alain C., and K. Wayne Smith. *How Much Is Enough? Shaping the Defense Program, 1961–1969.* New York: Harper & Row, 1971.

Fish, Lt. Gen. H. M. "The Security Assistance Program." *Air Force Policy Letter for Commanders, Supplement,* January 1976, pp. 18–22.

Fox, Annette Baker. *The Politics of Attraction.* New York: Columbia University Press, 1977.

Frank, Lewis A. *The Arms Trade in International Relations.* New York: Praeger, 1969.

Gaugash, Bill T. "Iran: Strategy for Survival." *International Relations,* Spring 1978, pp. 42–58.

Gelb, Leslie H. "Arms Sales." *Foreign Policy,* Winter 1976/77, pp. 3–23.

Gessert, Robert A., and William W. Cover. *Qualitative Constraints on Conventional Armaments.* McLean, Va.: General Research Corporation, 1976.

Gibert, Stephen P., and Wynfred Joshua. *Guns and Rubles: Soviet Aid Diplomacy in Neutral Asia.* New York: American-Asian Educational Exchange, 1970.

———. "Implications of the Nixon Doctrine for Military Aid Policy." *Orbis,* Fall 1972, pp. 660–681.

———. "Soviet-American Military Aid Competition in the Third World." *Orbis,* Winter 1970, pp. 1117–1137.

Golan, Matti. *The Secret Conversations of Henry Kissinger: Step-by-Step Diplomacy in the Middle East.* Trans. Ruth Geyra Stern and Sol Stern. New York: Quadrangle, 1976.

Haftendorn, Helga. "The Proliferation of Conventional Arms." In International Institute for Strategic Studies, *The Diffusion of Power: I. Proliferation of Force.* Adelphi Paper no. 133. London, 1977.

Harkavy, Robert E. *The Arms Trade and International Systems.* Cambridge: Ballinger, 1975.

Heikal, Mohamed. *The Road to Ramadan.* London: Collins, 1975.

Hoagland, Jim, and J. P. Smith. "Saudi Arabia and the United States: Security and Interdependence." *Survival,* March/April 1978, pp. 80–83.

Hoagland, John H. "Arms in the Developing World." *Orbis,* Spring 1968, pp. 167–184.

Hoffmann, Stanley. "The Hell of Good Intentions." *Foreign Policy,* Winter 1977/78, pp. 3–26.

Hout, Marvin J. "Munitions Export Control Policies and Procedures." *Defense Management Journal,* July 1970, pp. 49–52.

International Institute for Strategic Studies. *The Military Balance, 1972–1973,* and later volumes. London, 1972 and subsequent years.

———. *Strategic Survey, 1969.* Also volumes for 1970 through 1977. London, 1970 and subsequent years.

Iran, Imperial Embassy of [to the United States]. "Economic Development of Iran, 1945–1978." *Profile of Iran,* June/July 1978, entire issue.

Irani, Robert G. *US National Security Connections with Iran and Saudi Arabia.* Military Issues Research Memorandum. Carlisle Barracks: Strategic Studies Institute, U.S. Army War College, 1977.

———. "US Strategic Interests in Iran and Saudi Arabia," *Parameters* 7, no. 4 (1977): 21–31.

Joshua, Wynfred, and Stephen P. Gibert. *Arms for the Third World: Soviet Military Aid Diplomacy.* Baltimore: Johns Hopkins Press, 1969.

Kalb, Marvin, and Bernard Kalb. *Kissinger.* Boston: Little, Brown, 1974.

Kaplan, Stephen S. "U.S. Arms Transfers to Latin America, 1945–1974: Rational Strategy, Bureaucratic Politics, and Executive Parameters." *International Studies Quarterly,* December 1975, pp. 399–431.

Kemp, Geoffrey. "The International Arms Trade: Supplier, Recipient, and Arms Control Perspectives." *Political Quarterly,* October–December 1971, pp. 376–389.

Kennedy, Edward M. "The Persian Gulf: Arms Race or Arms Control?" *Foreign Affairs,* October 1975, pp. 14–35

Kissinger, Henry A. "The 1976 Alastair Buchan Memorial Lecture." *Survival,* September/October 1976, pp. 194–203.

———. "Reflections on American Diplomacy." *Foreign Affairs,* October 1956. Reprinted in *Fifty Years of Foreign Affairs,* ed. Hamilton Fish Armstrong, pp. 256–275. New York: Praeger, 1972.

———. *White House Years.* Boston: Little, Brown, 1979.

Krapels, Edward N. *Oil and Security: Problems and Prospects of Importing Countries.* Adelphi Paper no. 136 . London: International Institute for Strategic Studies, 1977.

Laird, Melvin R. *The Nixon Doctrine.* Washington, D.C.: American Enterprise Institute, 1972.

Laquer, Walter. "The Psychology of Appeasement." *Commentary,* October 1978, pp. 45–50.

Lefever, Ernest W. *TV and National Defense: An Analysis of CBS News, 1972–1973.* Boston, Va.: Institute for American Strategy, 1974.

Lenczowski, George, ed. *Iran under the Pahlavis.* Stanford: Hoover Institution Press, 1978.

Ligon, Walter B. "Foreign Military Sales." *National Defense,* July/August 1975, pp. 30–32.

Luck, Edward C. "The Arms Trade." *Proceedings of the Academy of Political Science* 32, 4 (1977): 170–183.

———. "The Soviet Union and Conventional Arms Control." *Proceedings of the Academy of Political Science* 33, 1 (1978): 57–65.

Lynch, John E. "Analysis for Military Assistance." *Military Review,* May 1968, pp. 41–49.

Mangold, Peter. "The Soviet Record in the Middle East." *Survival,* May/June 1978, pp. 98–104.

Mathias, Sen. Charles McC. *Mathias Report.* Newsletter to constituents. Washington, D.C., Fall 1978.

Meir, Golda. *My Life.* New York: Putnam's, 1975.

Morris, Beau. "Why France's Arms Exports Make It a Paper Tiger." *Armed Forces Journal International,* October 1978, pp. 19 et passim.

Nixon, Richard M. "Asia after Vietnam." *Foreign Affairs,* October 1967.

———. *Foreign Assistance for the Seventies.* Washington, D.C.: White House Press Office, 15 September 1970.

———. *Reorganization of U.S. Foreign Assistance Programs.* Washington, D.C.: White House Press Office, 21 April 1971.

———. *RN: The Memoirs of Richard Nixon.* New York: Grosset & Dunlap, 1978.

———. *U.S. Foreign Policy for the 1970's: A New Strategy for Peace.* Washington, D.C.: The White House, 18 February 1970.

———. *U.S. Foreign Policy for the 1970's: Building for Peace.* Washington, D.C.: Government Printing Office, 25 February 1971.

———. *U.S. Foreign Policy for the 1970's: Shaping a Durable Peace.* Washington, D.C.: Government Printing Office, 3 May 1973.

———. *U.S. Foreign Policy for the 1970's: The Emerging Structure of Peace.* Washington, D.C.: Government Printing Office, 9 February 1972.

Nutting, Anthony. *The Arabs: A Narrative History from Mohammed to the Present.* New York: Potter, 1964.

Ohio Arms Control Seminar. *Summary of Proceedings: Workshop II.* Columbus: Ohio State University, 1978.

Powell, Craig. "Arms Sales Is More than Just a Military Operation." *Armed Forces Management,* January 1968.

Pranger, Robert J. *Toward a Realistic Military Assistance Program.* Washington, D.C.: American Enterprise Institute, 1974.

Quandt, William B. "Soviet Policy in the October Middle East War —II," *International Affairs,* October 1977, pp. 587–603.

Ravenal, Earl C. "Toward Zero Military Assistance." *Hearing before the Committee on Foreign Relations, United States Senate, on S. 1443 to Authorize Furnishing of Defense Articles and Services to Foreign Countries and International Organizations.* Washington, D.C.: 2, 3, 4 and 8 May 1973. Reprint of extracted testimony.

Rivlin, Paul. "The Burden of Israel's Defence." *Survival,* July/August 1978, pp. 146–154.

Ross, Dennis. "Rethinking Soviet Strategic Policy: Inputs and Implications." *Journal of Strategic Studies,* May 1978, pp. 3–30.

Rubinstein, Alvin Z. *Red Star on the Nile: The Soviet-Egyptian Influence Relationship since the June War.* Princeton: Princeton University Press, 1977.

Sadat, Anwar. "Interview." *As-Sayyad* (Lebanon), 30 December 1976. Reprinted in *Survival,* March/April 1977, pp. 80–81.

Schandler, Herbert Y. "U.S. Arms Export Policies: A Candid Assessment." Address presented at seminar on "Foreign Military Sales and U.S. Policy," Center for Strategic and International Studies, Georgetown University, Washington, D.C., 19 September 1978.

Sheehan, Edward. *The Arabs, Israelis, and Kissinger.* New York: Reader's Digest Press, 1976.

Shulman, Marshall D. "Arms Control in an International Context." *Daedalus,* Summer 1975, pp. 53–61.

Sisco, Joseph J., ed. *Prospects for Peace in the Middle East.* Washington, D.C.: American Enterprise Institute, 1977.

Stanley, John, and Maurice Pearton. *The International Trade in Arms.* New York: Praeger, 1972.

Stern, Thomas. "Department Discusses Policy on the Sale of U.S. Military Articles and Services." *Department of State Bulletin,* 21 July 1975, pp. 98–102.

Stockholm International Peace Research Institute. *The Arms Trade with the Third World.* Middlesex: Penguin, 1975.

————. *Arms Uncontrolled.* Cambridge: Harvard University Press, 1975.

Szulc, Tad. *The Illusion of Peace: Foreign Policy in the Nixon Years.* New York: Viking, 1978.

Thayer, George. *The War Business.* New York: Simon & Schuster, 1969.

Treverton, Gregory F. *Latin America in World Politics: The Next Decade.* Adelphi Paper no. 137. London: International Institute for Strategic Studies, 1977.

United Nations. *The United Nations and Disarmament, 1945–1970.* New York, 1970.

U.S. Arms Control and Disarmament Agency. *Arms Control and Disarmament Agreements, 1959–1972.* Washington, D.C., 1 June 1972.

———. *The International Transfer of Conventional Arms: A Report to the Congress.* Washington, D.C.: Government Printing Office, 1974.

———. *World Military Expenditures and Arms Transfers, 1967–1976.* Washington, D.C.: Government Printing Office, July 1978.

U.S. Central Intelligence Agency. *Changing Patterns in Soviet-LDC Trade, 1976–77: A Research Paper.* Washington, D.C.: National Foreign Assessment Center, May 1978. (Unclassified publication.)

U.S. Congress, Senate, Committee on Foreign Relations. *Arms Sales and Foreign Policy.* Staff study. 90th Cong., 1st sess., 1967. Washington, D.C., Government Printing Office, 25 January 1967.

U.S. Defense Security Assistance Agency. *Foreign Military Sales and Military Assistance Facts.* Washington, D.C., December 1978 and prior years.

U.S. Department of State. *United States Foreign Policy 1969–1970: A Report of the Secretary of State.* Washington, D.C.: Government Printing Office, March 1971.

U.S. Military Academy. *Arms Transfers: A Senior Conference, 1976.* West Point, 1976.

Wattenberg, Ben J. "Is There a Crisis of Spirit in the West?" *Public Opinion,* May/June 1978.

Watts, William, and Lloyd A. Free. *State of the Nation, III.* Lexington, Mass.: Lexington Books, 1978.

Whetten, Lawrence L. *The Arab-Israeli Dispute: Greater Power Behaviour.* Adelphi Paper no. 128. London: International Institute for Strategic Studies, 1977.

Wood, Robert Jefferson."Military Assistance and the Nixon Doctrine." *Orbis,* Spring 1971, pp. 247–274.

Yankelovich, Daniel. "Cautious Internationalism: A Changing Mood toward U.S. Foreign Policy." *Public Opinion,* March/April 1978, pp. 12–16.

Index